Recycle Everything
Why We Must, How We Can

by Janet Unruh

Preface

This book describes a system for recycling everything we make—cars, clothing, computers and electronics, furniture, even buildings—a system for *material* sustainability. The book illustrates how the system can work and identifies new entities and roles necessary to a system of perpetual recycling. It also highlights the challenges and invites participation in the discovery of solutions.

Note: This book was originally published in May, 2009 under the title, *Whole-system Sustainability*.

Table of Contents

The Problem

Walk around any mega store, the kind where shelves of products tower far above reach, and just look at all the products – TVs, computers, refrigerators, clothes, and all kinds of gadgets. All these products have a useful life. After the first use, they may go to a second-hand store, or maybe not, and then they will be discarded. Imagine all the merchandise that passes through all the mega stores in a year; add the packaging and all the equipment used for manufacturing and transport. When all that stuff has reached the end of its life, where will it all go? To growing junk heaps all over this planet, on land and in the oceans. Most of us are aware of this by now. And yet, that's only the *downstream* view.

What's the *upstream* view? Where are the materials coming from to make all these products? Can these materials keep on coming day after day, year after year? Are supplies inexhaustible, or could they run out? The truth is, we're already bumping into limits.

Picture this: Somewhere at this very moment, there's a production manager for a large manufacturer, who's asking his materials coordinator about some parts that are overdue. Production is backed up while they wait for these parts. The materials coordinator says that parts suppliers can't get the parts they need either. It backs up all the way to a shortage of gallium, or indium, or terephthalic acid, or some other raw material.

Our ability to manufacture products is threatened by the increasing scarcity of raw materials. In some cases, raw materials aren't scarce per se, but they are becoming more expensive to extract because the sources that were easily accessible have already been scraped clean or siphoned off and remaining deposits are increasingly difficult to reach. The ore grades of these deposits may also be lower quality, in which case they will require a greater amount of energy to produce a ton of metal[1]. Access to materials can also be overshadowed by political tensions, regional conflicts, and war.

We only need to turn to the news to learn about shortages. An article in *New Scientist*[2] magazine reports that scientists are beginning to realize that certain raw materials will run out, according to a study conducted by researchers at Yale University, the U.S. Geological Survey, and the University of Augsburg, Germany. Here are a few examples. Indium, which is used in flat-panel TVs and computer screens, could run out in four to 13 years. Silver could become depleted in between four to 29 years. Lead, used in batteries, could become impossible to find in eight to 42 years. Zinc could be exhausted by 2037; indium and hafnium, which are important in computer chips, by 2017, and terbium, which is used to make fluorescent light bulbs, by 2012. Clearly, some of these shortages will have an impact on manufacturing soon.

New technologies often increase demand for certain materials. A few years ago, simultaneous growth in three industries – electronics, aerospace, and power generation – created a stampede for tantalum. Its price more than tripled. The semiconductor industry uses tantalum as a barrier metal, titanium and cobalt in transistors, and helium in wafer cooling. In electronics, the need for tantalum in capacitors surged with the growth in cell phones and personal digital assistants (PDAs). At the same time, demand increased in the aerospace industries, and power plants, which use super alloys that require tantalum. Demand from these industries converged and caused a shortage on the market[3].

In many cases, a substitute can be found, but sometimes there is a shortage of a material for which there *are* no substitutes. One example is piezoelectric materials, which generate electric current when they are under stress. Piezoelectric materials are used in the production and detection of sound, the generation of high voltages, electronic-frequency generation, scientific instruments, and ultra-fine focusing of optical assemblies. One kind of piezoelectric material is lead, and a critical use of lead is in batteries. Currently global stocks of lead are low. Just 40,000 tonnes (a tonne is equal to 110.25% of a U.S. ton) was recently available on the London Metal Exchange compared with 180,000 tonnes three years earlier. In addition, market data from the Shanghai Metal Exchange showed that, of six lead suppliers in China, four were out of stock in 2008[4].

Agriculture provides food, wood, and cloth materials for human consumption and use. But agriculture has serious problems too. The cost of nitrogen, which is critical to fertilization, is escalating sharply[5]. And there are many other stressors, such as farmlands in India becoming saline, wells running dry in China, and aquifers being drained in the U.S. Topsoil has washed or blown away, and mega farms are dosed with potent chemicals that provide artificial life support to crops. Entire populations living in arid parts of the world are dependent on imported grain. These practices are not sustainable over the long term.

Thomas Robert Malthus foresaw this predicament in 1798. In his *Essay on the Principle of Population*, he concluded that the world's human population increases at an exponential rate while food supplies grow at an 'arithmetic' rate. Thus, the need will inevitably exceed food supplies, and a large part of humanity will be doomed to famine. Malthusian theory predicts that humanity will enjoy increasing prosperity up to a peak in time when Earth's resources begin to run out. The peak will be followed by an unrelenting economic decline and mass starvation until the population decreases to a level that can be sustained by subsistence living.

In recent times, the Club of Rome arrived at the same conclusion about a future economic peak and decline. The Club of Rome was founded in April 1968 by Aurelio Peccei, an Italian industrialist, and Alexander King, a Scottish scientist. They invited a small international group of people from the fields of academia, civil society, diplomacy, and industry, to meet at a villa in Rome, Italy, hence the name[6]. Club members came together to discuss the issue of 'prevailing short-term thinking in international affairs and, in particular, the concerns regarding unlimited resource consumption in an increasingly interdependent world'[7].

Club researchers conducted a study to determine whether the productive capacity of the Earth could support the projected growth in human consumption. The researchers collected data and used a computer modeling system to determine their findings. In 1972, they issued the report, *Limits to Growth*[8], which sold 12 million copies in more than 30 translations. The report stated that if human consumption continued to grow at an exponential rate, humans would exhaust the resources of the earth and *overshoot* its capacity to support us. In 2004, the authors made improvements in the data, computer modeling, and testing. When they reviewed the updated results, they found that their previous estimates from 1972 were right on track. They published their findings in a new report, *The Limits to Growth: The 30-Year Update*[9].

Optimists predict that supplies will continue to flow because new technologies will remove obstacles to extraction. This is simply not realistic. Right now, there are those who are searching the entire planet for deposits of valued resources of all types. And yes, they are using powerful new technologies to expose and extract the resources of the earth. But these activities will come to a grinding halt when the difficulty and expense of extraction surpass the value of the resources on the market and subsidies fail to make up the difference. Yes, we've had some reprieves, and we may have more. But if we think that we have access to unlimited amounts of raw materials, we're in denial[10].

There's more information on the issues of availability of materials in the Appendix, including research conducted by the National Research Council on critical minerals and the

U.S. economy, charts showing the peak in prices of metals over the past 10 years, U.S. dependence on foreign sources of materials, and the stockpiles of critical materials maintained by the U.S. Department of Defense.

These studies and reports offer strong evidence of material scarcity. We'll abbreviate that discussion so we may move on. Assume that material availability is not guaranteed. Now we are faced with a real problem – what if we can't get the raw materials we need to continue the production and consumption of the many products that enhance our quality of life? *Could* there be some other way? The answer is *yes*. This book describes a vision of new system, one that is sustainable indefinitely. First, let's look at our current linear production-consumption system.

The Linear System

Right now, we are taking what we want from the planet, using it, and throwing it away. It's a one-way street from source to trash heap. *But if we think there can't be any other way, we are lacking imagination.*

Here is a depiction of the current linear system (following):

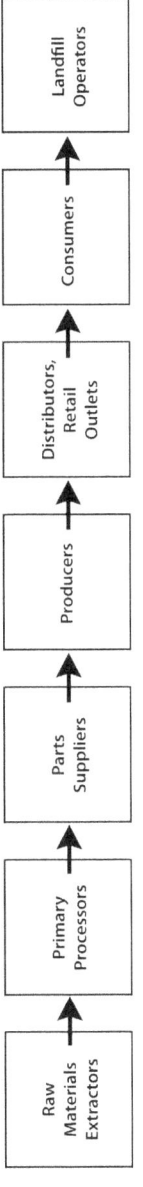

Figure 1: the linear production-consumption system.

First, we have the **raw materials extractor** – the companies that mine the ores and extract the chemicals, metals, wood, water, and petroleum. Crops also extract nutrients from the soil, so we must include agriculture.

Next we have the **primary processor**, who may purify the materials and create portions such as ingots of silver, spools of copper wire, and tanks of solvent.

Then we have **parts and subassemblies suppliers**, who create computer chips, radiators, wheels, windows, bolts of cloth, and countless items that are used to make finished products.

Then we have **producers** (manufacturers) who actually make the finished products, such as computers, modular homes, clothing, and appliances.

Next we have **distributors and retail outlets**. They sell finished products to consumers.

Finally, **consumers** use the products until they break down or until consumers buy a better model.

Eventually all these products go to **landfills**. Only a tiny fraction of the entire mass of products that flow through this system are recycled.

Materials are going into burgeoning junk heaps at the end of this process and at the beginning of it, manufacturers are searching for new sources of materials. *It's time to think about how we can connect the end to the beginning.*

Look around you. The microwave oven, the cell phone, the car, clothing, furniture – imagine that when you're done with them, you give them to a collector, possibly through curb-side pick-up. Then all these products would go back into the system for 100% reuse in new products. There would be no more used stuff being thrown on top of mountains of junk with toxic chemicals draining into rivers. What are the issues that need to be addressed for 100% reuse to become a reality? Why don't we fix the production-consumption system so that all the materials that enter into the system *stay* in the system? We must answer these questions. Only then can we say that our production-consumption system is sustainable. This is what we will explore in this book.

Imagine that you're holding a glass of wine. The glass you hold in your hand may someday become a window in a building. After that, it may become eyeglasses for someone or a coffee pot. Combined with other glass, it will have thousands of uses stretching into the distant future. Now imagine that you're driving your car. When your car has served its last day, the metal may go to become another car, part of a building, a cooking pot, a tool, or countless other things. This is how we should look at materials. They must become *perpetually reusable*.

Recycling Everything and Sustainability

Lots of people are talking about sustainability these days and there are many definitions. Unfortunately most of them are too vague and it is difficult to imagine how they could be put into operation. A popular definition of sustainability is 'meeting the needs of the present without compromising the ability of future generations to meet their own needs'. In the broadest sense, 'sustainability' refers to the ability (or inability) to sustain human practices long-term, e.g., the sustainability of energy production or of economic growth. But for this discussion, *sustainability is the ability to continue production and consumption indefinitely within the limitations of a finite Earth.*

This book describes a system for material sustainability, a continuous cycle of materials through production, consumption, and recovery in a never-ending cycle. The system for material sustainability has no dependence upon materials outside of itself nor does it allow materials to escape. The system continuously resupplies itself with materials that are within it. Once materials have entered the system, they never leave it.

Sustainability is often confused with environmentalism. Though related, *environmentalism* is the concept that the restoration and preservation of nature is most important, above all other considerations, whereas *sustainability*, as we just said, is the ability to continue production and consumption indefinitely within the limitations of a finite Earth. Sustainability means adopting a self-perpetuating, cyclical, closed-loop approach to production and consumption.

Downcycling and Blended Recycling

Another point that often confuses people is the distinction between *recycling* and *downcycling*. For instance, plastic gallon milk containers can be shredded and reused to insulate jackets or to make park benches; used tires can be made into doormats; wood scrap and plastic can be made into deck planks. These efforts are positive, but after the second or third use of these materials – then what? They go to the landfill. The value of these materials is then lost. This is *downcycling*, not *recycling*. What has been achieved is simply another use prior to disposal. The system is still linear, and not truly sustainable.

If the milk containers were *really* recycled, they would be melted down and reprocessed into *more* milk containers. The materials in tires would be recovered to make new tires. Materials from used, broken-down appliances would be used to make new appliances. Wood scrap and plastic would not be combined as it is now to make deck planks, but instead, the wood scrap would be composted to nourish the soil for tree farms and the plastic reprocessed and reused. Then we will have a perpetual, sustainable cycle.

Blended recycling refers to materials that have some recycled content, such as paper or steel—some kinds of steel are 80% recycled steel and 20% new. The higher the recycled content, the better; but we must set our sights on 100% recycled content.

You may have heard of eco-efficiency or life cycle analysis. These are methods used to identify, analyze, and manage the negative impacts of industry upon the environment. Methods such as these are beneficial, but they don't *stop* the flow of materials from quarry to dump. They slow down the rate of consumption and reduce the amount of materials flowing through the system, thereby succeeding only in *delaying the inevitable* depletion of resources.

Very few writers and professionals have begun to talk about connecting the end of current one-way system to its beginning, bringing the products back into the system at the end of their life, and using their materials again in a perpetual cycle. Among them are William McDonough, Michael Braungart of McDonough Braungart Design

Chemistry[11] and Dr. Robert Ayers of INSEAD[12]. More innovators, inventors, and creative engineers are needed with the courage and imagination to wrestle with these issues.

The Risk to Industry

All it takes is a threat to one vital material for an industry to be vulnerable to scarcity. It's obvious that continued production requires the availability of materials, and that producers should be aware of the current state of material availability.

A good risk assessment for producers would list every material used in the manufacture of their products, name their sources (vendors) along with alternative sources, assess the prospects for continuing supply, and estimate projected costs. Other important factors would be competition for those materials and projected demand for emerging products. The final result would be a materials list ranked in terms of risk and ease of mitigation. This would give producers the ability to see the areas in which they are vulnerable and prioritize steps they need to take to become sustainable.

In the next section, we'll look at the vision of how we can recycle everything.

The Vision

Is there an alternative to the linear system? Yes. Here's what we need to envision: that all materials used in production and consumption exist within a system, an inescapable system – a cyclical system in which they are perpetually recovered, reprocessed, and reused. In such a system, there is no such thing as 'disposal' of materials. They are owned, tracked, valued, and recycled *indefinitely*.

From here on, we'll refer to the system for material sustainability as *s4ms*.

Let's go on a tour of what an *s4ms* would be like. Figure 2 (following) is a drawing of a simplified, generic system for material sustainability. Notice that it is circular and that there is no entrance or exit for materials. Materials move around inside this circle among the roles or members of the *s4ms*.

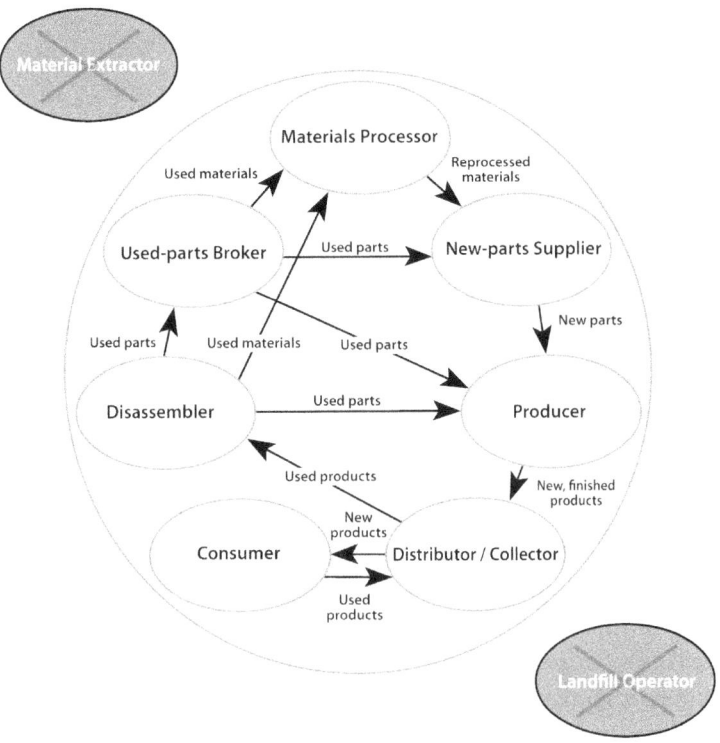

Figure 2: A simplified, generic system for material sustainability (s4ms).

Here we see that the raw materials extractor and the landfill operator are no longer in the system. However, there are two relatively new roles: the disassembler and the used-parts broker. The retail distributor takes on new responsibilities for recovery and collection of products from consumers. In fact, all the roles must adapt in order to exist in an *s4ms**.

Let's look at each role in depth, starting with the materials processor.

---

* System for material sustainability

The Roles

Materials Processors

Since an *s4ms** is not linear, there is really no starting point. However, we'll begin with material processors – this is the point at which materials return to their 'raw' or 'elemental' state after having been used in products. In an *s4ms*, material processors no longer receive raw materials from extractors. Material processors receive used parts and materials from used-parts brokers and disassemblers and reprocess them for a variety of uses. Ideally, the parts and materials are made of a single material, such as steel plates, plastic items, and glass bottles. Material processors return the parts and materials to their pure elemental state by a wide variety of means – melting them down, vibrating them with sonic waves, or shattering them with freezing temperatures or other methods. Material processors then create portions, such as spools of pure copper wire, ingots of steel, sheets of glass, blocks of dry ice, or containers of pure chemicals. Many materials that are currently in use will have to be reformulated to permit reprocessing. The materials processors sell the purified and processed materials to new parts suppliers or they provide reprocessing services for a fee.

Some materials may need purification. In times past, heating to white-hot liquidity materials such as metals or glass purified them, and impurities were allowed to burn off. However, in an *s4ms*, the resulting smoke must be trapped and put to use somehow. It would be better to invent a technology that separates all the materials and to have a specific kind of reuse in mind for each one – *including* the 'impurities'.

---

* System for material sustainability

14

Some materials processors process agricultural materials. These processors take agricultural 'waste' and wood from furniture and other sources and grind them up for compost. Materials processors also process fruits and vegetables to make juices and pulp and mill grain to make flour.

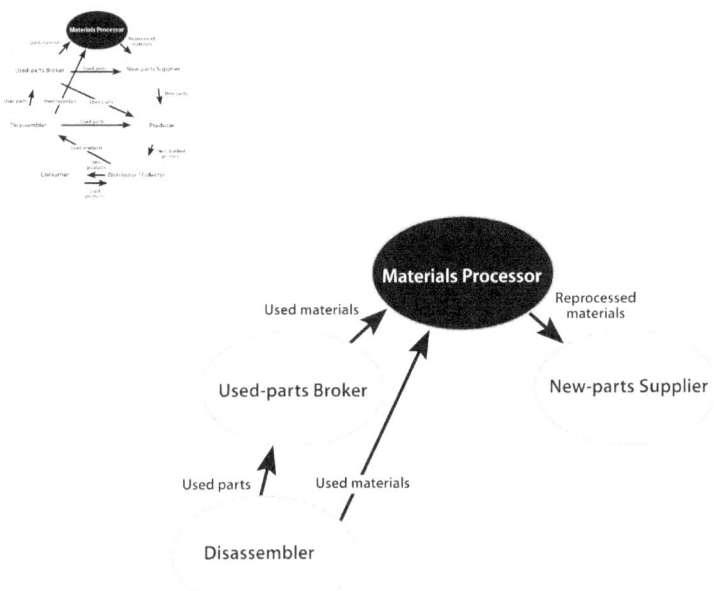

Figure 3: The relationship of materials processors to other roles in the system.

New-Parts Suppliers

New-parts suppliers create new parts from reprocessed materials. These new parts become components of finished products. New-parts suppliers buy materials (blocks of steel, fruit juice, or sheets of glass) from materials processors to create the components they sell to producers. Components can be things like rope netting, windows, bolts, insulated wire, doorknob assemblies, plastic bottles, cans, ink, paper, or tomato sauce. The producers assemble these components with others to make a finished new product of some type.

From here on, when we say 'parts' we're including 'assemblies' which can be everything from a door hinge to diesel engine, which is installed into something larger to create a finished product.

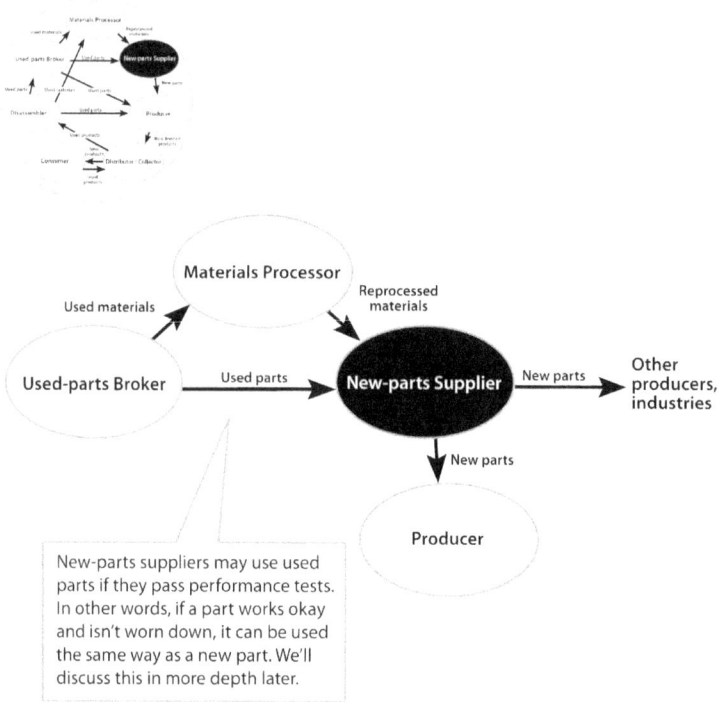

Figure 4: The relationship of new-parts supplier to other roles in the system.

Producers

Producers manufacture products. Producers make new, finished products from new parts purchased from new-parts suppliers or used parts purchased from used-parts brokers and obtained from disassemblers. Products include TVs, cars, trucks, clothing, books, toilets, frozen food, toys, electronics, carpet, and millions of other things.

In an *s4ms**, it is essential for each product to have an RFID (radio frequency identification) tag. The tag is used to track the materials as they move around the system. The tag can only communicate with a specific reader used by the distributor/collector or the disassembler, so consumers don't need to worry about privacy issues. The tag carries information about the product, including its serial number, material makeup, disassembly, testing and reprocessing instructions. The producer sends new, finished products to the distributor to lease or rent (not sell) to consumers. We'll talk about ownership later.

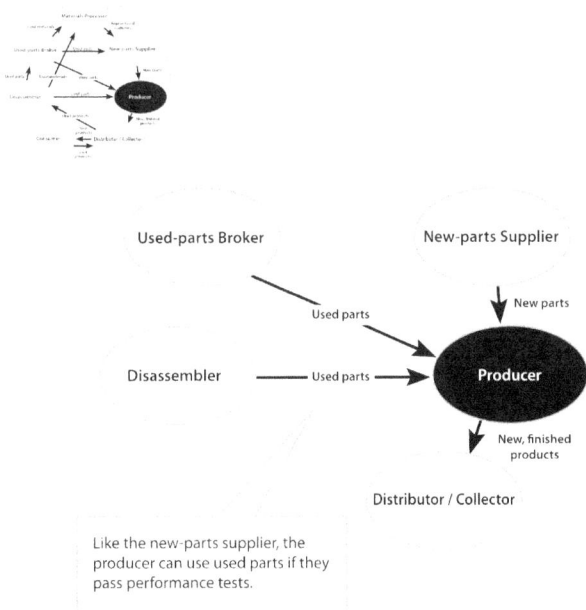

Figure 5: The relationship of the producer to other roles in the system.

---

* System for material sustainability

Distributors/Collectors

Distributors/collectors serve a two-way function. They provide an outlet for new, finished products to rent or lease to consumers, and they serve as a collection center for used products retrieved from consumers. The RFID tag helps automate inventory control for both new and used products. Depending on the agreement with the consumer, the distributor/collector may provide maintenance and repair service, exchange new models for old ones, or pick up old models from the consumer. Transportation of products and in-home services are specified in the agreement between the distributor/collector and the consumer. The consumer may also bring the old model back to the collector. If the consumer disposes of the product, a penalty is assessed. The distributor tests returning products and may rent or lease them to second-tier consumers. If used products are no longer serviceable, the collector sends used products to the disassembler.

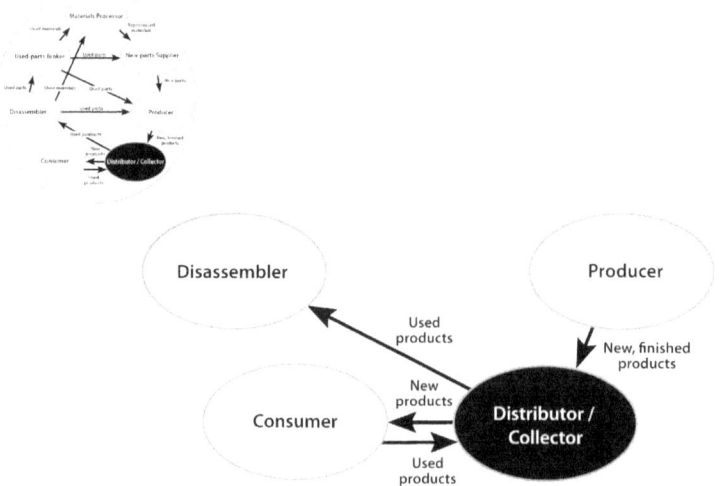

Figure 6: The relationship of the distributors/collectors to other roles in the system.

Consumers

Consumers use new products. Consumers rent or lease new products from the distributor and depending on the agreement, return used products to the collector or trade them in for new models. Consumers pay a monthly fee to rent or lease the product. Fees may be automatically debited from consumers' bank accounts. Consumers are financially responsible for any damage that may have occurred while the product was in their possession and for disposal – which is not permitted in the *s4ms**.

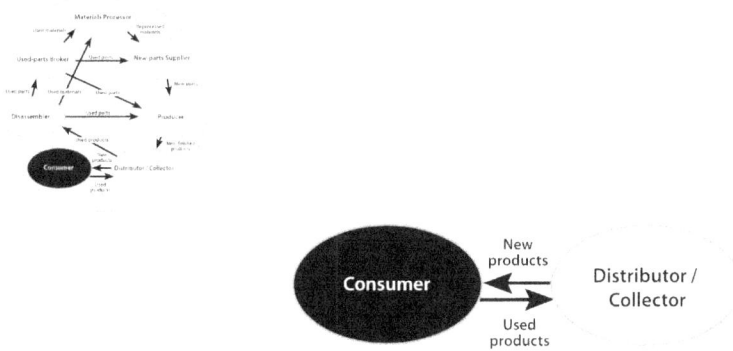

Figure 7: The relationship of consumers to other roles in the system.

Disassemblers

Disassemblers take used products apart. Disassemblers receive used products from distributors/collectors. The RFID tag in each product reports its product information (model, make, serial number; and material content), and handling instructions (disassembly, test and reprocessing), and one more thing we haven't mentioned yet – the condition of the product obtained from condition sensors. Disassemblers check the condition of the parts and decide whether they can be used again.

---

* System for material sustainability

Disassemblers read the disassembly instructions from the RFID tag and take used products apart according to the producer's disassembly plans. For each product, there is a specific plan and method for taking it apart. The disassembler works with the producer to select equipment and computer systems that meet the needs of the producer's disassembly plan. For example, a robot may squeeze an electronic music player so that it pops apart into several pieces. Another robot might desolder it. Naturally, the disassembler must have robots that can do this work. All of this is determined in advance through collaboration with the product designer/manufacturing engineer, who writes the disassembly plan.

Disassembly could be centralized in a disassembly factory designed for many different products made by different producers. Producers could also have their own disassembly factories to guard proprietary secrets.

After disassembly, the disassembler may send parts back to the producer, or sell them to a parts broker or to a materials processor, depending on the agreement with the producer and the condition of the parts.

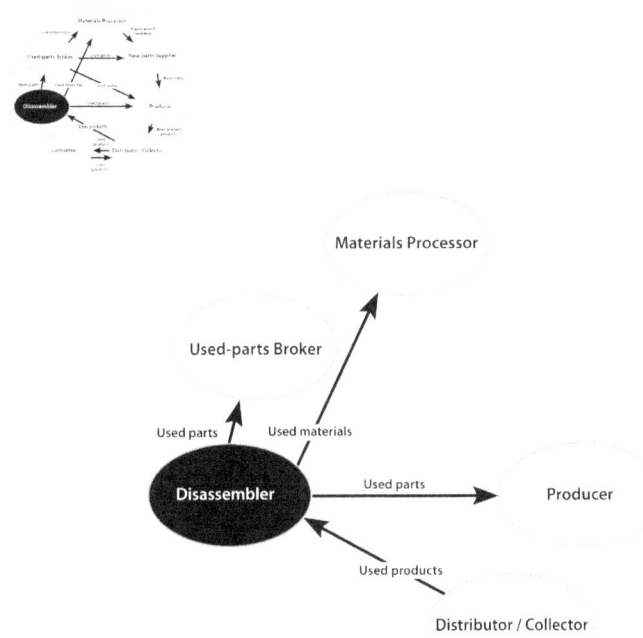

Figure 8: The relationship of disassemblers to other roles in the system.

Used-Parts Brokers

Used-parts brokers buy and sell used parts. They receive used parts from disassemblers and sell them to new-parts suppliers and producers. The used-parts broker buys and sells all types of parts. If the used-parts broker has accumulated a quantity of a particular type of part, made of a single material, the parts broker may sell the whole quantity to the materials processor for reprocessing. The demand for the materials may be greater than the demand for the part.

There are currently regulatory restrictions regarding the sale and use of used parts as new. This will take some re-thinking and some new definitions as used parts take their places in new products. Maybe products can be 'new' but with 'remanufactured' parts. Xerox has already been using remanufactured parts in new copiers for years, as we'll see later.

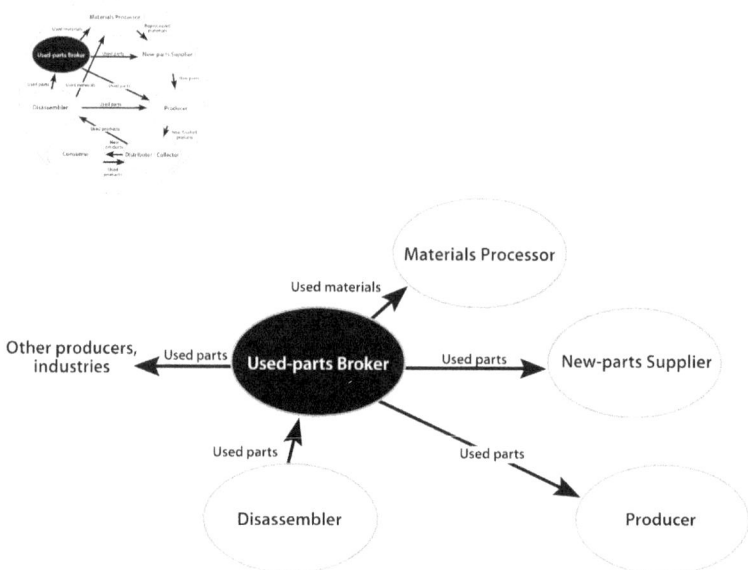

Figure 9: The relationship of used-parts brokers to other roles in the system.

Now we have seen each role in a fair amount of detail. There are many things that will be explained in more depth later. Here is the system in generic form (following):

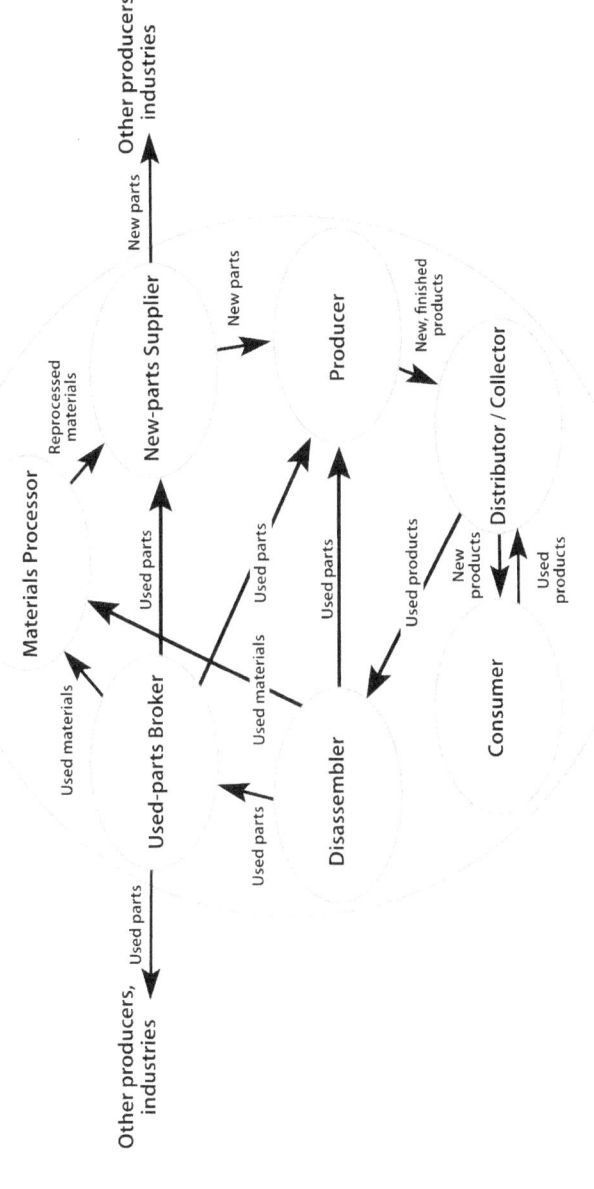

Figure 10: A generic system for material sustainability (s4ms).

An *s4ms** can work for products such as TVs, microwave ovens, and laptops – in fact, all durable and 'soft' goods, all consumer and business-to-business products[13].

Let's take an example. Suppose that you, the consumer, rent a TV. When you are ready, you can either return it to the collector or trade it in for a newer model from the distributor. In an *s4ms*, the old TV now goes to a disassembly factory that takes it apart and sends the materials and parts back to the producer or to a used-parts broker. The producer may make a 'new' TV out of the 'old' TV's parts and materials or the broker may sell the parts to other producers, even other industries. The materials *never* fall out of the system. This is what makes it sustainable.

Examples of Systems for Material Sustainability

Cars

Here's another example. Suppose that your car can no longer be used, and it doesn't make sense to repair it. Depending on your agreement, you or the distributor/ collector transport the car to the distributor/collector's service center. The distributor/collector tests it and if possible, repairs it and leases it to someone else, or has it transported to the disassembler.

Your old car is not going to be flattened with all its parts broken, intermingled, and impossible to recover. It comes into the disassembly factory and immediately, the RFID tag communicates with the factory's computer system. The tag tells the computer system that the condition of the car's major components is poor and that the car requires a particular disassembly process.

---

* System for material sustainability

Let's say this process calls for the paint to be blasted off with frozen CO_2 pellets and captured for reuse. Then robots with grippers and lasers cut through welds and rivets to separate the major body parts, such as the roof, hood, trunk lid, doors and body. Other robots separate the engine, transmission, exhaust system, various components, panels, wiring, hoses, and controls.

All the parts go through triage and are sorted into categories – reuse as-is, repair, or reprocess. The disassembler's computer system communicates with the producer's computer system to find out whether the producer wants to retain ownership of parts that can be reused as-is or repaired and reused. If so, the parts are sent back to the producer.

The disassembler sends parts that the producer no longer wants to the used-parts broker. The used-parts broker serves as a clearinghouse of used car parts for all interested parties. The used-parts broker sorts parts, repairs them as needed, and lists them in an online database to make them available for purchase by materials processors, new-parts suppliers, and producers. Complex parts, like an electronic subassembly, must be tested and guaranteed by the used-parts broker.

If parts are no longer usable or marketable, the disassembler or the used-parts broker sends them to the materials processor. Some materials are harder to reprocess than others. Copper, if pure, is relatively easy to process because it can be melted down and formed into wire or plates for use in all sorts of industries. But if a materials processor receives a complex part like a circuit board printed with metallic ink, the producer has to supply instructions on how to separate the materials. The materials engineer writes the plan for processing the materials and identifies the equipment needed, and the producer works out an agreement with the materials processor for processing specialized or complex materials. After the materials processor reprocesses the materials back into pure elements, he/she sends them to the new-parts suppliers who make new parts. From there, the materials return to the producer.

Back to the new car example: new-parts suppliers and used-parts brokers send components for new cars to the producer. Consumers come to the car dealership (the distributor/collector) and lease a new car. If they wait a month or two, there's a chance that their new car may have some parts and materials from the old one!

Here's how the system for material sustainability would look for cars (following):

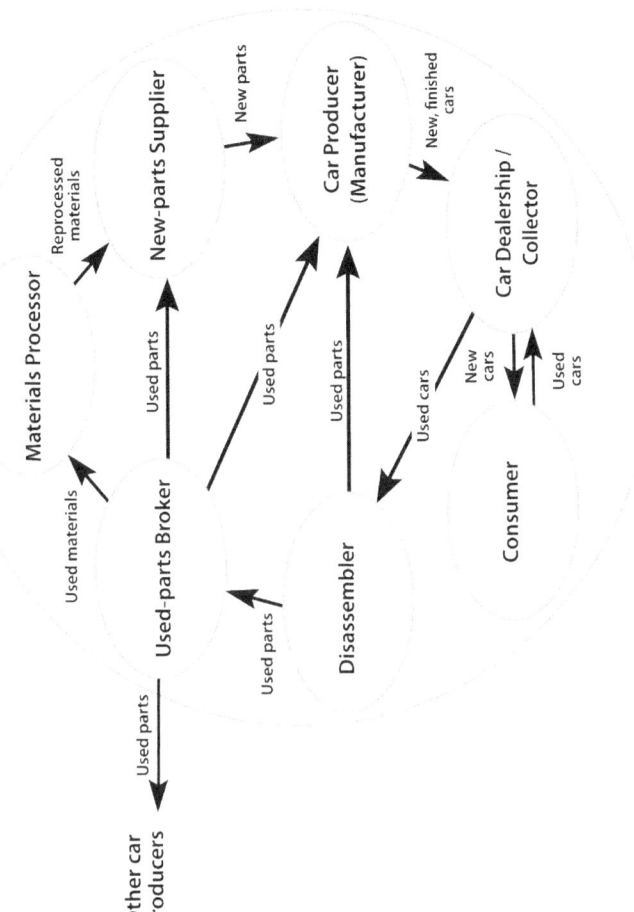

Figure 11: A system for material sustainability (s4ms) for cars.

The *s4ms** for cars that we just discussed is an example of a wholly inorganic system. For things such as metals, minerals, glass, plastic, and chemicals – the *inorganic* materials – the goal is to recover, reprocess, and reuse them so that they never lose their value and are perpetually reusable. In an *s4ms*, materials such as these are no longer

---

* System for material sustainability

extracted, which means producers have to become independent of raw materials.

But there *is* a limited role for the raw materials producers in cases where the uses of organic raw materials are well justified. Most of us associate the word *organic* with organic food. But in this context, *organic* refers to biological sources of materials as opposed to *inorganic* materials, mentioned above. *Organic* raw material producers are farms, logging and fishing operations, orchards, vineyards, meat and dairy producers. In the organic materials system, the ideal is *not* to cut out the raw-materials producer as it is in the inorganic materials system. In fact, organic materials (cotton cloth, wood furniture, agricultural waste, etc.) are perpetually reusable. The difference is that after they have been used, they are routed to compost and then to fields to support new crops.

Soft Drinks

Let's take a look at an example of an *s4ms** that brings in organic materials – soft drinks in plastic bottles. Flavorings could come from fresh fruit processed for flavoring. The producers of soft drinks will also need water, CO_2 (for carbonation), plastic for bottles, and paint for labeling.

So let's say that the materials processor takes raw lemons and limes and processes them into juice. Another materials processor processes plastic and paint that have been recycled. A new-parts supplier uses the plastic to make bottles and another new-parts supplier takes the water and CO_2 to make carbonated water and the fruit juice to make flavorings. The producer paints the bottles, mixes the carbonated water and flavorings and fills the bottles. The soft drinks are then distributed to consumers.

---

* System for material sustainability

After consumers imbibe the soft drink, they recycle the bottle at a collection center. The bottle goes to the disassembler, who recovers the paint from the bottles using the ether that evaporated from the paint when the producer painted the bottles (we'll discuss this kind of process in more depth later). The materials processor reprocesses the bottles into plastic, which new parts suppliers use to create new bottles for new soft drinks. The cycle is complete.

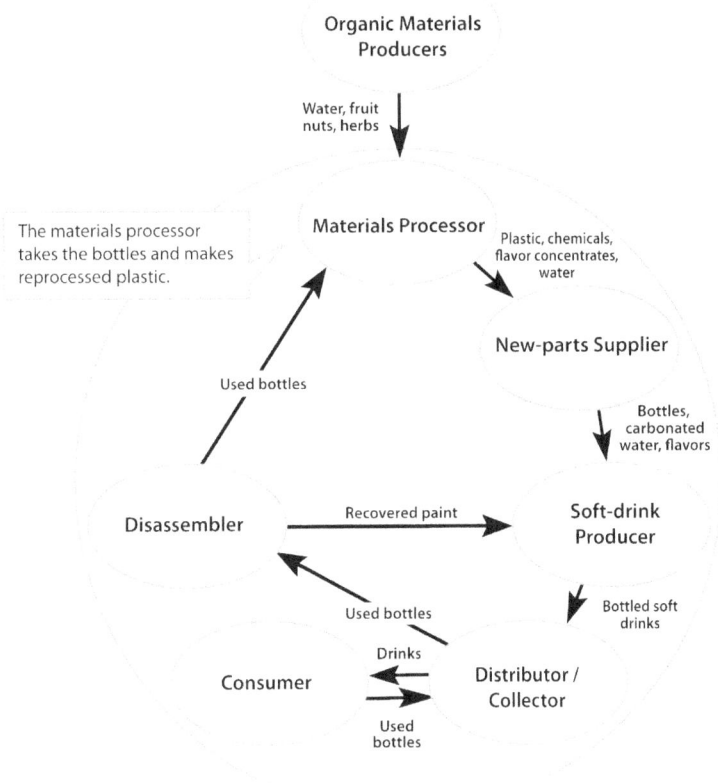

Figure 12: An organic system for material sustainability (s4ms) for soft drinks.

Obviously, many products are made of organic sources such as wood, cotton, other natural fibers and food. Although the goal of the inorganic materials is to be recyclable indefinitely, the goal for organic materials is to be composted. Organic materials can be *downcycled* through multiple uses before becoming the stuff of soil enhancement.

Blue Jeans

Let's look at another example that is almost 100% organic: blue jeans.

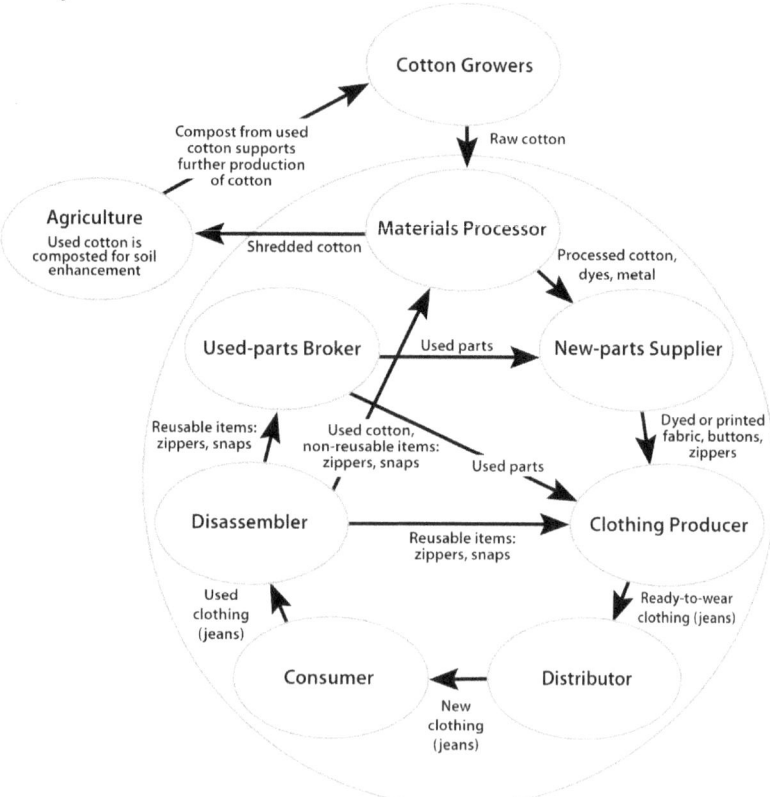

Figure 13: An organic system for material sustainability (s4ms) for blue jeans.

Here we see that the raw materials producers provide cotton, dyes, and metals. These materials are processed into fabrics, snaps, and zippers. Then these parts are made into jeans. The clothing producer sends the blue jeans to distributors who in turn, sell them to consumers.

Clothing may be the sort of thing that doesn't need to be returned to a distributor/collector but can go from the consumer directly to the disassembler. Since a pair of blue jeans may pass through a couple of owners, it doesn't make sense to track them. But old clothes still have a value – a small one – so there is a role for someone to collect them and sell them to the disassembler. Second-hand stores could do this. The disassembler separates the cotton from zippers and snaps and sends parts to the materials processor. The materials processor melts down the metals and shreds the used cotton for use as compost. The shredded cotton contributes bulk and nutrients to the soil so that it can produce another crop of cotton. Thus, the circle is closed.

An important consideration in the organic system is *sustainable yield*. Here, the principle is this: *the rate of harvest is limited by the rate of regrowth*. In the past, some business owners have made it their goal to liquidate an entire resource, such as a forest, and reap the profit even if it means closing the business afterward. A sustainable use of a resource avoids boom and bust. Though supply flows may be constrained by the sustainable yield, demand is assured. The company that depends on that resource can continue to exist and be profitable into the indefinite future.

Lumber

Let's look at logging. For the sake of argument, let's say it takes an average of 25 years for a marketable kind of tree to grow to maturity. To make the math simple, we'll also suppose that the logging company has rights to 25,000 acres. That means the company could log and replant one 1000-acre patch of forest per year. The patches would be sequenced so that each year, the company would log and replant a patch that's a mature 25 years old.

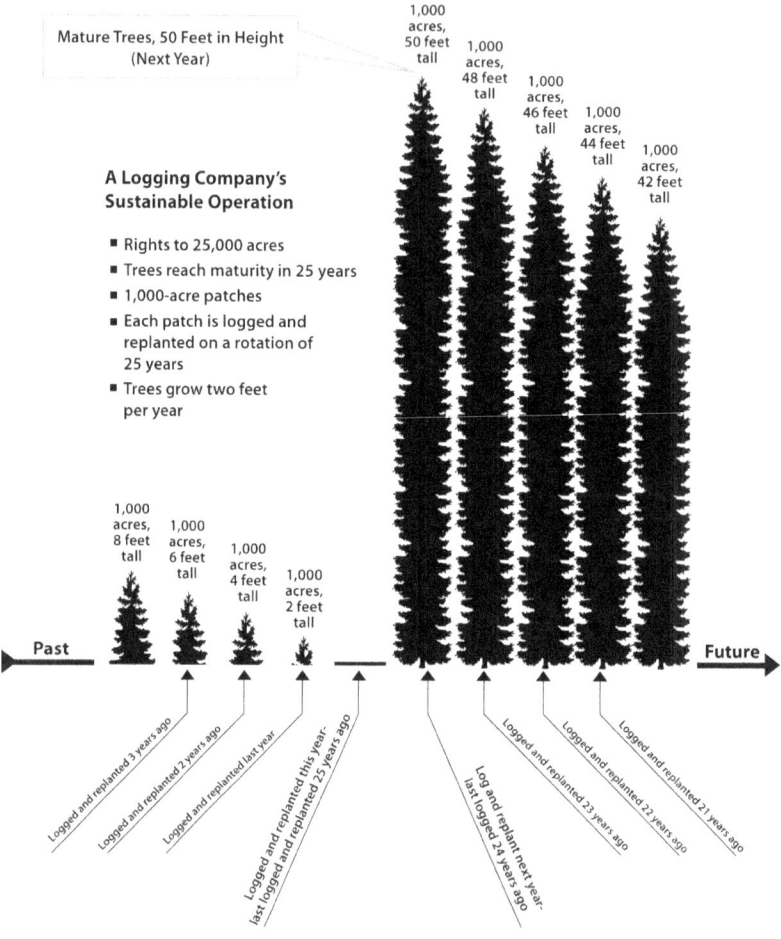

Figure 14: An organic system for material sustainability (s4ms) for lumber.

People are cutting down forests much faster than the rate at which forests can regrow. We need to dramatically reduce our use of wood, not just because the supply is dwindling, and not just because entire species are vanishing, but because the forest performs an important function. Forests control global warming by absorbing carbon from the atmosphere. They also provide oxygen for us to breathe. There are too many trivial and shortsighted uses of wood. When a hurricane advances on a city, people board up their windows with plywood. After the storm has passed, they discard the plywood. It's ironic to think that for purposes such as these, forests are cut down that otherwise would aid in controlling such storms.

Construction is the biggest consumer of materials and generates the most waste by far. In the future, buildings should be built with disassembly and reuse of materials in mind. Instead of using wood for construction, plastic should be used. Plastic can be formed into sheets and boards of all shapes and sizes. When it has served one purpose, the plastic can be recovered, reprocessed, and reused. Wood can be used for hardwood floors, cabinets, and furniture as long as it is varnished with a biodegradable finish. Even this type of use will become a luxury in the not-too-distant future. If the natural beauty of wood is to be painted over, then producers should use plastic instead. Window frames, doors, decks, furniture – anything that is currently made of wood could be made of plastic, a perpetually-reusable material. If we human beings succeed in replacing fossil fuels as a means of generating energy, the petrochemical industry's next big product could be plastic. We're going to need a lot of plastic to replace most of our uses of wood.

Let's talk about biodegradable plastics since it may seem like a good idea for plastic to be biodegradable. Commercial bio-based plastics including PLA (polylactic acid), PHB (polyhydroxybutyrate), PHV (polyhydroxyvalerate) are made of starch-based and soy-based polymers. Previous attempts to create biodegradable plastics resulted in plastics that only broke up into smaller pieces of plastic. But the new organic-based polymers are biodegradable back to elemental carbon. This may sound like good news at first but here's the issue that emerges for biodegradable plastics: the production of plastics from agricultural sources will compete with food production as we saw recently with the use of corn for ethanol. Even the so-called agricultural 'waste' should be used to replenish nutrients to the farmland soil rather than being made into ethanol or plastic. Fertile soil is made of plants that have decomposed and built up rich topsoil. If all the 'waste' is scraped off the land, it will not support crops in any way except artificially. And that means using petroleum-based fertilizers – another ironic misuse of resources.

Tilling the soil, as human beings have done for millennia, causes valuable topsoil to blow or wash away. But researchers are making progress in the development of perennial crops, including wheat. Wes Jackson, founder and president of the Land Institute in Salina, Kansas[14], wheat breeder Steve Jones[15], and plant pathologist Tim Murray of Washington State University[16] are developing perennial wheat. When this type of wheat is fully developed, it will eliminate the need for tilling. Although not a new idea, perennial wheat is attracting increasing attention from researchers and farmers.

The domesticated prairie is also not tilled, and it takes the idea of perennial crops a step further. In the domesticated prairie, perennial wheat and other grains, legumes, and oil-producing seeds are grown and harvested together, and their output is sorted afterward. This type of field takes advantage of natural benefits of the prairie, which maintain a healthy diversity that eliminates the need for pesticides and herbicides. Perennial crops have demonstrated high yields.[17]

Industries that harvest a wild catch have greater difficulty in determining how much they can take without endangering the supply long-term. The salmon industry and other fishing industries are currently struggling to ascertain their sustainable yield. Eventually, consumers may have to eat less salmon and pay more for it, but that's better than not having any.

There have been so many technological and scientific breakthroughs in the past few decades that it hardly seems necessary to point out that many things have been accomplished that were formerly unthinkable. There are probably some aspects of the idea of recycling everything that seem unthinkable now. But we can't let ourselves get caught up in the technical challenges of *how* we're going to reach our envisioned goal. First, we need to be clear about what the goal *is*.

Four Areas of Innovation – Overview

Now that we've seen several systems for material sustainability, we need to look at four areas of innovation that are needed to make these systems work. These are the four areas:

Materials Engineering

Materials must be engineered to be reusable for an indefinite number of times. We have to start with materials because right now, the materials that make up most products we see today are intended for one use only. They can't be recovered, reprocessed, and reused without enormous effort and expense and most likely, not at all. But imagine products that are designed to be taken apart, the parts melted down, shredded or in some way transformed, and returned to the stockpiles. To accomplish this, producers must *engineer* materials to be reused.

Product Design / Manufacturing Engineering

In the world of manufacturing, products are designed to be easy to assemble at the factory. But in an *s4ms**, products also have to be easy to *disassemble*. Most products are built with rivets and welding that are not designed to be taken apart. This needs to change. If fasteners are easy to unfasten (snaps, screws, puzzle-piece assembly), disassembly can simply be assembly-in-reverse, but in other cases, disassembly may involve laser-cutting, freeze-shattering or other means. Whatever the method, it must be easy and cost-effective to take products apart. And it must produce parts and materials that can be reused, repaired or reprocessed.

Materials Management

Because producers need materials to make products, they must *own* the materials in the *s4ms*. Producers track their materials as they travel around the system using a combination of RFID (radio frequency identification) technology and an online database.

Revenue Model

In an *s4ms*, producers no longer sell a product; they lease a property to their consumers. The entire system operates on the revenue generated in this way. Except for the consumer, the roles in the system bring in revenue through transactions with each other.

Let's look at each of these areas in more detail.

---

* System for material sustainability

Materials Engineering

To be sustainable, we need to back up all the way to the materials that we use to make products. It doesn't make sense to recover used products, disassemble them, and hope to reuse their materials if the old parts can't be made into new parts. Until now, we have put products together without any thought of how their materials could be recovered. There hasn't been a plan to take the TV, microwave oven, or laptop computer apart and reuse the materials to make another, perhaps better TV, microwave oven or laptop. But imagine that the materials that make up these products could, by various means, be wadded up like balls of putty and thrown back into a giant stockpile, ready to be used again. But to do that, materials have to be *engineered to be reusable*. This is the greatest challenge for recycling everything and the one to tackle first.

For the sake of discussion, let's define materials as the nontoxic elements of the periodic table plus water, wood, biomass, glass. This includes all forms – solid, liquid, or gaseous. Plastics and alloys are included *if* they can be recovered, reprocessed and reused indefinitely. Although we may prefer to avoid toxic materials for environmental reasons, the point here is that materials that degrade easily should be avoided.

The new vision for materials is *perpetual reusability*. Such materials are reusable an indefinite number of times, ideally without loss, degradation, or useless byproducts generated in reprocessing.

If materials can't be recovered and reused again for their original purposes, it is generally because they have been *mongrelized* – mixed with other materials in such a way that results in a total loss of recovery value after use.

This doesn't necessarily mean that materials can't be combined. However, all combination materials must either be perpetually reusable or their constituents have to be separable into their raw-material forms. Materials can't be mixed with other materials irreversibly, if that mixture results in a loss of recovery value. A metal alloy, for example, combines materials to attain the desired qualities such as strength, resistance to corrosion or flexibility. If it can be reprocessed and reused without having to add more of something and without generating waste, it qualifies as a perpetually reusable material. But a substance like vinyl can't be reused except perhaps in a downcycled form of filler material.

Consequently, when materials engineers develop a new combination material, they must also devise a cost-effective method for reprocessing it. If the material is not perpetually reusable, a reprocessing method will be needed to separate its constituents into their pure state to allow easy reuse for future products in multiple industries. The materials engineer must think it through and describe ways in which materials could be reused by other industries.

Examples of Perpetually Reusable Materials

There's already been some progress in creating, recovering, reprocessing, and reusing materials. Let's look at some examples that show that materials engineering of this kind is possible.

Plastic Bottles – Dupont

Soft-drink and water bottles are made of polyethylene terephthalate (PET), which is a kind of polyester[18]. In the United States and Canada over four billion pounds of PET are produced each year, half of which is used in soft drink bottles. PET is currently made from petrochemicals[19]. However, PET can be made from ... PET.

DuPont Corporation created a process that can use as inputs all types of polyester films, fibers, and plastics with high levels of contaminants, such as dyes and metals, even blended cotton-polyester fabrics, and reprocess them into PET. The process is called the Polyester Regeneration Technology or the Petretec^{SM} process[20]. The Petretec^{SM} process uses chemical reactions to essentially 'unzip' the polyester molecules into their constituent monomers. The tremendous advantage of this is that these monomers are identical to those used as raw materials for the creation of polyester. Because of this, there are no limits on the reuse of PET. This reduces the use of petrochemicals and avoids disposal of plastics in landfills or by incineration. The monomers are FDA approved for all uses including food containers.

Engines – Recieder

In Europe, new regulations are forcing producers to find ways to recycle almost everything. Between 17 and 20 million cars, trucks, and buses reach the end of their useful lives every year in the European Union (EU). Disposal generates around 17 million tonnes of waste, which is typically contaminated with heavy metals. Vehicles that have reached the end of their useful lives are called ELVs, End-of-Life Vehicles. The ELV directive requires all metal components from ELVs to be separated and decontaminated for reuse.

A joint venture called Recieder (Zaragoza, Spain) has developed and demonstrated a new high-capacity process using advanced technology in an automated disassembly plant to break down and separate the metals from used engines for reuse.

The process does not involve chemical reactions; it uses a combination of physical separation techniques. The process cuts the motor into small pieces and separates all the metal materials using magnetic and current separators and other technologies.

The use and combination of these technologies into an automated plant designed for the separation of metals has been a first. And it has achieved impressive results. *The different metal alloys from an ELV engine – iron, aluminum, and heavy metals – were separated with 99% efficiency.* The recovered metals were of high quality. In particular, producers plan to use the aluminum in the manufacture of new engines and transmission parts. The recovered ferrous metals are more than 99% pure and have a high value as a resource for foundries[21].

Carpet – Shaw Industries Group and InterfaceFLOR

Carpet is another example of a material that is recycled and remade into the original product – carpet. Shaw Industries Group, Inc., makes nylon carpet through a process of recycling used carpet. The company is able to reuse recycled carpet repeatedly without any loss of aesthetic or performance properties. Most commercial and residential carpet is made of type 6 nylon (nylon 6). Shaw established a nylon recycling operation in Augusta, Georgia, which converts this type of used nylon carpet back to its original material and provides nylon for new carpet. The company works with recycling companies throughout the U.S. to have used carpet collected and delivered to its recycling plant[22].

Even more impressive are recent developments at Interface, Inc. The company has been collaborating with Universal Fibers, Inc. for 10 years to develop a process that can reclaim all types of carpet, commercial and residential, regardless of the type of fiber or backing used, such as nylon 6, polyester, or polypropylene. This effectively eliminates the need to dispose of *any* type of carpet into landfills or by incineration in the future. Interface's subsidiary, InterfaceFLOR, recovers carpet and recycles it for use in its products. The company expects to divert 30 million pounds of carpet from landfills annually with this new technology.

Prior to this, Interface diverted tens of millions of pounds of carpet from landfills solely by recovering carpet in places where the Interface flooring products were being installed. With the new breakthrough in recycling and reusing carpet, InterfaceFLOR can reclaim carpet from any source[23].

Office Machines – Xerox

A decade ago, in 1998, Xerox Corporation introduced its first fully digitized copier, the Document Center 265 Digital Copier, which was more than 90% remanufacturable and 97% recyclable. The copier had only about 200 parts whereas conventional copiers had more than 1,250. Its sales exceeded forecasts. That same year, remanufacturing and waste reduction saved Xerox $250 million[24]. This product came very close to realizing the company's ambitious goal of 'zero to landfill for the sake of our children'. The machine made extensive use of replaceable assemblies and parts that could be used again and again. In one year, Xerox prevented more than 145 million points of material from entering landfills through equipment remanufacture and parts reuse and recycling. Xerox diverted almost 9.5 million pounds of additional waste from landfills through return programs for Xerox supplies such as cartridges[25].

Xerox already had an existing capacity to reuse and remanufacture its machines. Product lines had been designed in an integrated fashion so that some of the components were interchangeable between different machines on the assembly line but the system was less than optimal. This changed when Xerox's Environmental Leadership Program, which began in 1990, explicitly required products to be designed for reuse, remanufacturing, and recycling.

As a result, business processes were altered dramatically. Among the processes that were reengineered were the return of equipment, assemblies, and parts to central warehouses, control of inventory, including scrapped parts, and the reuse of equipment, assemblies, and parts in remanufacturing. Xerox developed a triage for its assets as follows: reuse as-is, repair, reprocess, recycle, or dispose of by landfill or incineration.

Since 1991, Xerox's efforts in recycling and remanufacturing have enabled them to refurbish more than 2.8 million copiers, printers and multifunction products[26]. Returned products that are suitable for reuse undergo rigorous testing before remanufacturing. Those that cannot be remanufactured are disassembled, and the parts are reused or recycled. A small fraction of the remaining material is discarded. In 2006 alone, Xerox collected 43,000 metric tons of equipment and reused or recycled 96% of it[27].

Xerox has developed the ability to forecast the return of products and rely on them as a source of components for new products. *The company attributes several hundred million dollars in cost savings each year to the reduction of raw material needed to produce new machines*[28].

These examples show that with some ingenuity, it is possible to create products whose materials can be recovered, reprocessed, and reused. The chances of being able to do this are greatly increased if the materials themselves are designed for it.

The New Vision For Materials

Let's focus on inorganic materials first. Clearly, we can't rely solely on pure, elemental materials; we need combination materials that have particular qualities needed in products. There are two values to keep in mind when designing new combination materials. One is that *materials must be perpetually recyclable*. The new combination materials should be stable materials that can be reused an indefinite number of times for multiple purposes by multiple industries. This is the top preferred type of material. The other value is that materials must be *lossless,* both in terms of degradation and market value.

If you recall, mongrelized materials are those that are mixed in such a way that they cannot be separated, resulting in a total loss of recovery value after use. No doubt, many producers rely on mongrelized materials and don't want to take on the cost of finding other materials that can be reused. But if they use mongrelized materials in their products, neither they nor anyone else will be able to use the materials more than once or twice, because their value will be lost.

A combination material must be either perpetually reusable as-is, or separable into its constituent parts. If a combination material can't be reused as-is and its constituents can't be separated, they should not be combined in the first place. The materials engineer needs to find a different approach. For example, the aluminum in soft drink cans is an alloy whose constituents can't be separated. To make more aluminum soft-drink cans, the recycled aluminum must have 'virgin' aluminum added to it. Because of this, aluminum soft-drink cans are *not* a perpetually recyclable material. The better approach would be PET plastic bottles because PET *is* a perpetually recyclable material.

Let's draw some distinctions among materials to clarify what is desirable or acceptable and what is not.

Class A Materials

Class A materials can be recovered, reprocessed and reused indefinitely. These materials are 'elemental', that is, pure elements defined by the periodic table of elements nontoxic elements of the periodic table plus water, wood, biomass and glass. This includes materials in all forms – solid, liquid, or gaseous. Plastics and alloys are included *if* they can be recovered, reprocessed and reused indefinitely with virtually no degradation.

Class A materials are easily bought and sold and are useful to many industries.

The process for returning these materials into their pure, elemental state after use should not generate byproducts that can't be used elsewhere. If a ton of clear glass is melted down, it is inevitable that small amounts of paint, soft drink residue, and a dead fly or two would get into the mix and be reduced to ash. The ash could be separated off the glass, and perhaps if the ash passes safety tests, it can be used for certain types of agriculture. At some point, we need to establish a maximum number of parts per quantity of material for residues of this type. If the material exceeds the maximum, then the material engineer needs to reexamine the materials and make adjustments. If the ash or other impurities are useful for some other form of production – that's all the better.

Class B Materials

These materials are not 'elemental,' but they are special-purpose combination materials that are perpetually reusable. Plastics, alloys and 'smart' materials (discussed later) are included in this category as long as they can be reused indefinitely and retain their value on the market.

Class C Materials

These are materials that have extensive reprocessing requirements and generate substantial amounts of waste but are reusable to some extent. If the process can make use of reusable chemicals, the materials could be considered Class C materials. But if a chemical reaction is needed for reprocessing the material and the byproduct is a large quantity of toxic liquids, this type of material would be a Class D material (discussed below). Class C materials also include those materials that must have quantities of other materials combined with them to regain their usefulness after recycling. An example may be drywall, which could be ground to a powder and remade into drywall with the addition of water and other chemicals. Similarly, aluminum alloys require more raw aluminum to regain their value after use.

Class C materials need to be redesigned to improve their recyclability, decrease waste, eliminate the need for additional materials, and reduce reprocessing costs. Better yet, they should be phased out when a Class A or B replacement is found. The most important distinction is this: if the market value of these materials is retained to some extent after the costs related to reprocessing and disposal, these materials are Class C materials. But if there is no market value after the first use, these materials are Class D materials.

Class D Materials

These are the mongrelized materials – irreversible combinations of materials with limited or no reusability. The value of these materials is lost after the first or second use. Class D materials may be downcycled as filler but ultimately they go to enlarge landfills. These materials need to be phased out completely.

Organic Materials

These materials are created from biological sources. Organic materials can be downcycled in a succession of lower-grade products until they become suitable for compost. For example, wood furniture may be downcycled to wooden tools and to wood chips for landscaping. The compost returns the value of the material to the field, where new organic materials are produced, such as natural fibers such as cotton, hemp, wood and of course, all types of food.

If a material such as wood is used for furniture, it must be treated only with biodegradable products. A piece of wood furniture that has been finished with a plastic coating is not suitable for grinding up and composting. Unless it can be demonstrated through trials that a synthetic chemical is non-toxic, it is preferable not to take a risk with human health and the environment as whole. Things that are part of the organic system should not be combined with things from the inorganic system unless there's an easy way to separate them at the product's end of life.

Recovery Challenges

The toughest areas for recovery may be in areas such as cleansers and drugs. After their use, their chemistry goes down the drain. The best option in terms of recovery would be to make wastewater facilities capable of filtering out everything but pure water molecules. Everything else – the drugs passed through urine, cleansers, even things that we may now believe to be harmless, such as organic fertilizer (taking a lesson from phosphates from many years ago) – should be removed from water before we put it back into the larger ecosystem. There is an argument for removing these chemicals from the water for the sake of the environment, and if they can't be recovered for reuse, then they should still be filtered out to avoid unintended effects in the ecosystem. Of course, the use of chemicals in agriculture is an entire discussion by itself and is out of scope for this writing. But if we are going to recycle everything, these chemicals also need to be recovered and reused. Imagine a technology that can sort out molecules in the public water system. We might be surprised at the things we could recover.

Summary – Materials Engineering

In the *s4ms**, the most highly valued materials are those that can be recovered, reprocessed, and reused perpetually without loss. Elemental materials retain their value on the market and combination materials (plastics, alloys) also retain their value *if* they are perpetually recyclable. Materials that are combined in such a way that they can't be reused and therefore lose their value are *mongrelized* materials. These should be phased out as quickly as possible.

Research and Development Directions for Materials Engineering

Here are some research and development challenges for materials engineers:

---

* System for material sustainability

- Begin examining the kinds of materials that comprise everyday consumer or business-to-business products. Collect requirements for these products. Can the materials used in these products be modified to become perpetually reusable materials? If not, what sorts of perpetually reusable materials need to be invented to fulfill product requirements?

- Resolve conflicts between reusability and desired qualities of materials. An example would be the rigidity of plastics and metals achieved through combination of materials that currently excludes reuse.

- Identify ways in which materials re-engineering forces change in product design and manufacturing engineering.

Examples of Materials Engineering Projects

The challenge for materials engineering is to convert materials that are in current use to perpetually-reusable materials or invent replacements.

These projects may be quite challenging. Remember, the goal is to take materials that are in currently in use and reformulate them so that they can be perpetually reusable. Here are nine projects to consider:

- Rigid plastics – used for TVs, stereos, laptops, office furniture
- Paint – used on metal, glass, plastic
- Non-stick coatings – used in cookware
- Rubber – used in tires
- Paper – used for business, publishing and personal needs – eliminate the need for new supplies of wood or replace with plastic

- Ink – used on paper – create a type of ink that can be removed and recovered
- Soft plastics – used for shopping bags, toys, sheet plastic, shrink-wrap
- Steel – used in construction, vehicles, cookware
- Cement – used in construction, streets and sidewalks

Later in this book we'll talk about the Institute for Material Sustainability, which is a non-profit organization that the author is working to establish. The purpose of the institute is to provide services to industries that will help them make the transition to systems for material sustainability. One of the first things the institute plans to do is to identify projects that will fulfill the need for perpetually reusable materials. The institute will challenge qualified research and development specialists in both the public and private sectors to take on projects. The institute will also assist in obtaining funding as needed. We'll talk about the institute and its goals later in this book.

Product Design / Manufacturing Engineering

First, let's define *product design* and *manufacturing engineering* for those who are not familiar with manufacturing. Product designers conceptualize a product that is aesthetically pleasing, functional, and that conforms to safety requirements and company image. They then design or select the parts to be used in the product and develop lists of parts called 'bills of materials'. Manufacturing engineers take the design and the parts and determine the best way to build the product at the factory, so that parts are put on in the correct order, labor is spread evenly on the assembly line, materials and tools are positioned conveniently, and processes are safe and efficient. Together, product designers and manufacturing engineers take the design from the 'drawing board' and work out a plan for building the product at the factory.

The goal of recycling everything puts new requirements upon product design and manufacturing engineering: to make disassembly and remanufacturing of products *easy* and *cost effective.* This means that product designers and manufacturing engineers must plan for the 'fate' of the product upfront. The outcomes are limited to a few possibilities. If a used product passes performance tests at the distribution / collection center, it can be leased again as-is. The re-leasing of products is only possible if they are designed for longevity. Product designers and manufacturing engineers must strive to design high-quality, long-lasting products; built-in obsolescence is no longer profitable. But when a product can no longer be re-leased, it is sent to the disassembler.

After the product is disassembled, there are three basic possibilities: used parts, especially electronics, must pass performance tests to find out if they can be *reused as-is*, or *repaired and reused*. If they can't be reused or repaired, then they must be completely disassembled and routed to the materials processor for *reprocessing.*

The product designer and manufacturing engineer have to ensure that disassembly and remanufacturing is cheaper than purchasing new parts. This is mostly accomplished by reusing parts in new products as much as possible, as we saw in the Xerox example earlier.

In an *s4ms**, product designers and manufacturing engineers use materials that are perpetually recyclable and lossless. This means that they may have to try something different from what they did in the past because of new constraints and opportunities posed by changes in available materials. Once the product is designed, the work isn't over yet. The manufacturing engineer is responsible for writing three kinds of plans for the product: a *disassembly plan*, a *test plan*, and a *materials reprocessing plan*. The materials reprocessing plan is based mostly on information from the materials engineer. Let's look at these three plans in more detail.

The disassembly plan is a specific plan for disassembly that is relatively easy and cost-effective to carry out and that recovers the maximum number of parts and assemblies for reuse in new products. The plan entails the selection of disassembly equipment, computer programs, processes, personnel and testing. The plan mostly likely involves instructions to be read by a computer system that controls specific programmable robots. If the disassembly plant is more low-tech, the instructions can be displayed on computer screens for workers. This plan must be the result of an agreement worked out between the manufacturing engineer and the disassembler.

---

* System for material sustainability

The second is a test plan that determines whether the parts can be reused or have to be reprocessed. This applies primarily to electronics or computer-related components but any kind of part that can deteriorate from use must be tested. This test is essentially the same as the performance tests that were used when the product was originally built. The used parts that pass the performance test are repaired if necessary and sent to the producer's assembly line for production.

The third plan that the manufacturing engineer has to create is a specific plan for reprocessing the materials used in parts that are damaged and can't be reused. The plan is based on specifications determined by the materials engineer. Like the disassembly plan, the reprocessing plan also must be the result of an agreement worked out between the manufacturing engineer and the materials processor. It will involve the selection of materials reprocessing equipment, computer programs, processes, personnel and testing – in this case to validate material content. The process must be as easy and uncomplicated as possible and the equipment needed should be available at the materials processor's plant, not something that is special or complex. This helps to reduce cost.

Each of these three plans – the disassembly plan, the test plan, and the reprocessing plan – is given a name or designation and stored in the product's RFID tag. When the product arrives at the disassembly plant or the material processing plant, the computer system recognizes the designation and reads the instructions for these three plans along with the producer, product name, serial number and material content. We'll look at RFID and information systems in more depth in a later section of this writing.

Disassembly Methods

Product designers and manufacturing engineers, let's get your thought processes going. Let's say you're designing a part with plastic-coated metal. How do you get the plastic off the metal? Can it be heated and *pulled* off? Can both materials be reprocessed and reused?

Or perhaps you're planning to use insulated copper wire in your product. If it passes performance tests, used wire may be reused *as is* in a new product. But the wire has been damaged, you'll have to have a way to separate the copper wire from its insulation. A high-speed stripper could be used to split the insulation and roll the copper into a coil. The copper could then be sent to the materials processor for reprocessing. The insulation also needs to be reprocessed and reused. You'll need to find a supplier that can provide insulated copper wire that is fully reusable.

How about painted glass? You'll need some way to get the paint off the glass. Some ways to do that could be blasting, peeling, or dissolving with solvents. Can the vapors that escape from the paint when it's drying be captured and recombined with paint taken from returning products to make 'new' paint? This might sound impossible at first, but there are companies that are doing such things.

There's much more to disassembly than getting the paint off the product. Product designers/manufacturing engineers already strive to improve products by reducing the number of parts and decreasing the number of steps to assemble products. The new twist is to reduce the number of steps to *disassemble* products. What are some ways that products can be assembled and disassembled? Of course, there's screwing and unscrewing, zipping and unzipping, and snapping and unsnapping. Hot or cold temperatures may be used for stretching, deforming, shattering, expanding, or shrinking materials to build a product or take it apart. Products could be made with pressure-point assembly and disassembly so that the disassembler would only have to press those points and the entire product would pop apart.

Laser cutting could sever welds, and metals could be separated with electromagnetism as we saw in the engine disassembly example earlier. Vibrations of various kinds – sonic, microwave, etc., may be used to break down complex molecules into simpler ones.

An exciting development for disassembly is 'smart' materials, particularly shape-memory materials. A shape-memory material is designed to hold a set shape until a trigger temperature is applied (outside the normal range), and then it changes to a second set shape. Trigger temperatures can be set quite accurately. Other means of triggering the disassembly may be through microwave, infrared, pressure, sound, computer and robotic control, electric current, or magnetic fields. Shape-memory materials can be polymers or metal alloys.

There are lots of videos of animated computer models on the Active Disassembly website[29]. The first example shows a cell phone with heat applied to it. The cell phone's screws elongate and lose their thread when heated, allowing the cell phone to fall apart. The cell-phone parts are then easily sorted out by size and weight into bins.

In another example, thermoplastic hot-melt adhesives are applied to layers of a flat-screen panel. When re-heated to the trigger temperature, the adhesive releases its grip and falls out from between the layers. The assembly is then easy to take apart. Another kind of shape memory join that is used for two parts looks like a zipper – two interlocking strips that hold each other tightly with interlocking capital-

'T' shapes. When heated, they relax into straight prongs and release the parts.

Maybe someday, every product will have a command module that is triggered at the disassembler's plant that causes the product to disassemble itself. It's an idea.

Disassembly Plans

Disassembly plans include both the disassembly process and the equipment required at the disassembly plant. As we have said, each product has an RFID tag that provides instructions on how to disassemble it and if needed, how to reprocess the materials. Here's a sample report that would be displayed on the computer screen from an RFID tag of a product that's coming in at the disassembler's factory:

> Hi, I'm a cell phone, made by Wow Corporation, model 103C^{30}, serial number 10-AGN2-2503. My electronics got wet and no longer function. Disassemble me using process RU202-A, which requires the gentle squeeze-pop robot number Y2670. You should be able to recover my case, dial and display glass. My electronics need to be replaced. Desolder the ELTT module using process RU220-B and equipment NPT-2097. Replace the ELTT module using assembly process AP02912. Send the used ELTT module to the material processor, who should use reprocessing method 209H, which uses two levels of heat to separate, detach, and reprocess parts.

The RFID tag tells the computer system at the disassembler's plant the quantity of every material contained in the cell phone, based on the producer and model information. The computer system enters these data on the materials into a database. This information will travel with the materials until they return to the producer or are sold.

The computer system begins the automated process of disassembling the cell phone according to the instructions given by its RFID tag. First the cell phone is routed on a conveyer belt to the squeeze-pop robot number Y2670. The robot grips the cell phone and presses on several pressure points, which are identified by process RU202-A. The case, display, keyboard, speaker, microphone, and electronic modules come apart. One of the electronic modules is routed to a desoldering gun (NPT-2097), and the ELTT module is detached using process RU220-B, which heats four tiny welds and sucks the welding material away from the module. The module is then routed to a shipping container for small parts that will go to the materials processor when enough parts are accumulated.

The other cell-phone parts are sent back to the producer. If the producer doesn't want them, the disassembler sells them to the used-parts broker.

Materials Reprocessing Plans

Materials reprocessing plans include both the processing plan and the equipment required. The materials processor receives shipping containers from many different disassemblers and accesses each database for information about the materials and their weight (or volume). The ELTT, an electronic module from the Wow cell phone, is one of the more difficult items to process because it has a plastic base and copper wiring. Some modules may have to be handled by a specialized disassembler.

For example, let's imagine that the ELTT module requires a special reprocessing method, a method that subjects the module to 100 degrees centigrade and separates copper wire from the plastic base. As it is heated, the plastic drips through a wire mesh, and is collected, purified, and reprocessed into sheets. The copper is routed to a vat where it is melted down, purified, and reprocessed into various portions or shapes. This can be cost-effective if several tens of thousands of modules are processed at one time.

The computer system at the material processor's plant receives the information about the materials from the disassembler and tracks the amounts of materials and the producers that own them in its database. When the materials are processed, those amounts are sent to the new-parts supplier to be made into new parts for the producers. The material processor is responsible to ensure that the quality of the materials meets established standards. This is the only point in the entire system that some material may be disposed of because it was generated as a byproduct or it was a waste residue and there is no use for it or it is toxic. The disposal method must neutralize any unstable elements. The ultimate ideal however, is zero waste.

Remanufacturing

Used parts come in to the producer from the disassembler and the used-parts broker. When used parts arrive, they must go through triage to sort the parts into different categories. There are two basic possibilities for reusable used parts: 1) those that can be reused as-is, or 2) those that can be repaired and reused. The disassembler and used-parts broker sends reusable parts to the producer; parts that can't be reused go to the materials processor.

Let's break that down. Suppose a consumer swaps an old cell phone for a new one at the distributor/collector's center. The distributor/collector tests the old cell phone and determines its condition or reads the condition from the RFID tag. If the cell phone is in bad condition, the distributor/collector sends it to the disassembler. The disassembler reads the data on the cell phone's condition as reported by the RFID tag and sorts its parts by their condition – okay as is, needs repair, or not usable. At this point, the disassembler asks the producer if he/she wants the parts. If the producer doesn't want the parts, they go to the used-parts broker, who repairs and resells them. If the producer wants the parts, the disassembler sends the parts that are okay as-is or in need of repair to the producer. The disassembler sends the worn or damaged parts to the materials processor for reprocessing, and from there they go to the new-parts supplier to be made into new parts and then back to the producer.

When used parts are returned to the producer, they go through a cleaning and testing process. When the parts in need of repair are ready for reuse, they are combined with the reuse-as-is parts and new parts. The part numbers are identical. *Used and new parts are handled in exactly the same way from this point on.* They go through a supply process from inventory to queues at the assembly line.

Summary – Product Design / Manufacturing Engineering

Material sustainability puts new requirements on product design and manufacturing engineering. Products must be designed for disassembly and remanufacturing. There is a wide range of disassembly methods available, but assembly and disassembly methods need to be integrated to make disassembly as easy and cost-effective as possible. This means that products must be built to be taken apart. RFID will aid the disassembler by providing information on the product, its condition, and the process and equipment needed for its disassembly. The disassembly process is automated. Parts are divided by a triage process into reuse as-is, repair and reuse, or reprocess. The manufacturing engineer provides plans for disassembly, testing, and materials reprocessing. When used parts return to the producer, they are repaired as necessary and then handled exactly the same way as new parts.

Research and Development Directions for Product Design and Manufacturing Engineering

Here are some challenges for product designers and manufacturing engineers to meet within a system for material sustainability.

- Collaborate on new approaches to designing products that use only Class A and B materials.
- Develop cost-effective disassembly methods.
- Design disassembly processes and equipment that are adaptable to a variety of products.
- Design a disassembly facility and a materials reprocessing facility for a given set of products.

Materials Management

Materials management has to do with the ownership of materials and consequently, the tracking of materials throughout the system. As we said earlier, once materials enter the system, they *stay* in the system. Who is ultimately responsible for materials? The producers are, because they need materials for their products.

For this reason, producers must *own* the materials needed for production. This is how producers achieve their independence from raw materials extractors. Materials, in the form of products, leave the producer's factories, cycle through the system and return. Ownership of materials is critical to an *s4ms**.

Business experts have promoted just-in-time processes, whereby the materials and parts to build products arrive at the moment that they are needed at the assembly plant. However, as shortages become more frequent, it will become apparent that it is better to stockpile needed materials and parts. The cost of storage is minor compared to the cost of downtime at the factory.

In this section, we're going to talk about how materials are tracked as they pass through the system and return to the producer. We will see how materials are handled when producers retain ownership or sell their materials. In the next section, Revenue Model, we will look at how each role in the *s4ms* earns income and pays expenses.

RFID

RFID is essential in tracking materials after they leave the producer. Some consumers are concerned that RFID will be used to monitor their activities and track their attendance at

---

* System for material sustainability

religious or political meetings. In an *s4ms**, there is no interest in consumers' affiliations or whereabouts. RFID is used only to provide this information:

- Producer (e.g., Love-it Corporation)
- Name of the product (e.g., Love-it T.V. / Computer Monitor)
- Model (e.g., model #09217)
- Serial number (e.g., 1097-1297)
- Material content (e.g., 1 kg. glass, 800 kg. polymer 56, 9 g. copper, etc.)
- Disassembly plan (e.g., disassembly plan #2057-2 using robot 982)
- Test plan (e.g., test plan #1-85184)
- Reprocessing plan (e.g., glass reprocessing plan #122098, polymer 56 reprocessing plan #10980-9, copper reprocessing plan #1-0975, etc.)
- Condition of the product (e.g., 'this product is functioning normally'.)

For this purpose, the 'semi-active' type of RFID tag is ideal. It is a battery-powered tag that collects data about the condition of the product from the product's sensors, but does not broadcast a signal. When the product comes within range of a reader at the distribution/collection center or the disassembly plant, the reader initiates the communication. No one else can activate the RFID tag.

The distribution/collection center has to keep information on file identifying consumers who bought a product because eventually the product has to be returned or retrieved. However, these data are already being stored in databases today when consumers buy products with their credit or debit cards. The only difference would be that cash purchasers would have to provide contact information.

---

* System for material sustainability

For the purpose of an *s4ms**, the RFID tag only needs to be read when:

- The producer sends the product to the distributor/collector.
- The distributor/collector receives the product.
- The distributor/collector leases it to the consumer.
- The consumer returns it to the distributor/collector.
- The distributor/collector ships it to the disassembler.
- The disassembler receives it.

The product information (the producer, name of the product, serial number, etc., mentioned above) is stored in an online computer database that is shared by all the roles in the *s4ms* except the consumer. As the product or its materials cycle through the system, the data moves with it.

Throughout the journey, the materials are tracked online. The producer puts an RFID tag on products going out to distributors/collectors. The RFID tag stays on the product until the disassembler's system reads it and accounts for all the materials and their weights or volumes. The disassembler removes the tag and sends it back to the producer. After the tag is removed, the computer system tracks parts to be reused by part number, and it tracks materials to be reprocessed by weight or volume in an online database.

The disassembler sends used parts to the producer, the used-parts broker or the materials processor depending on their condition and whether the producer wants them. If the parts are reusable, the disassembler sends them to the producer or to the used-parts broker and tracks them by part number. If the parts are to be reprocessed, they are tracked

---

* System for material sustainability

by weight and volume from the time they leave the disassembler and pass through the materials processor and the new-parts supplier and return to the producer.

The table below shows how the data about products and their materials are tracked as they pass from one role to the next.

FROM WHOM, TO WHOM	PRODUCTS / MATERIALS	TRACKING METHOD	COMPUTER INFORMATION
Producer to distributor/collector	Batch of new products	RFID / Online database	List of product serial numbers, each with its individual product information
Distributor/collector to consumers and consumers to distributor/collector	Individual new / used products	RFID	Individual product serial numbers matched to consumers who lease them
Distributor/collector to disassembler	Batch of used products	RFID	Individual product serial numbers, product information and condition
Disassembler to producer	Batch of parts that passed performance tests	Online database	Part number, quantity of parts batched as 'reuse as is' or 'repair'.
Disassembler to materials processor	Batch of parts that didn't pass performance tests	Online database	Name of the material and units of weight or volume and owner
Materials processor to new-parts supplier	Materials ready for production	Online database	Name of the material and units of weight or volume and owner
New parts supplier to producer	Batch of new parts	Online database	List of part numbers, material content and weight or volume and owner

Table 1: Materials tracking method.

What about proprietary secrets? Disassemblers, materials processors and new-parts suppliers must be entrusted to handle products that are proprietary with contract agreements. This is needed in cases where a new kind of material – perhaps a 'smart' material of some kind – is used. Parts that are obtained from a used-parts broker are nonproprietary because their producers are no longer protecting their content (design and material) from those who want to reverse-engineer them. The new-parts suppliers also make nonproprietary parts and may combine new parts with used ones from the used-parts broker. The sale of these nonproprietary parts includes the going price of the materials that are in them. Figure 15 (following) shows producer-owned materials and how they are tracked.

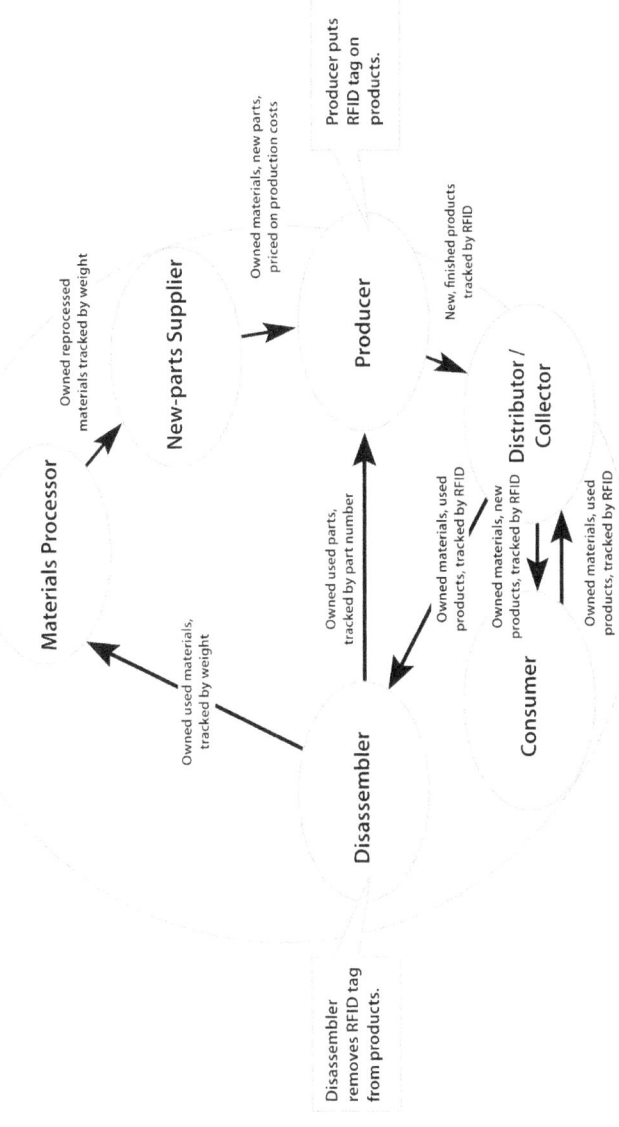

Figure 15: Producer-owned materials and how they are tracked.

All of the materials in the preceding system are producer-owned. The used-parts broker is left out because he/she does not have a role to play when the producer wants to retain ownership of the materials.

Let's say the product is a car. The disassembler takes apart cars that are no longer serviceable, and there are three possible destinations for the parts depending on whether they are to be reused, repaired or reprocessed.

Some parts, like light fixtures, window assemblies, windshields, and other non-critical parts may be reused as-is if they pass a performance test at the disassembler's plant. These go to the producer and are tracked by part number. Parts that fail the performance test are sent to the materials processor with their weight or volume information.

Some materials may be reprocessed because of new product designs – the body of the car, for instance. The disassembler sends the metal plates from the body of the car to the materials processor where they are melted down and purified. Paints and coatings are separated and recovered using a specified process. The resulting metal is then made into sheets, or whatever shape is convenient for the new-parts supplier, who then makes new metal plates, possibly using the recovered coatings as well. The new-parts supplier sends the coated metal plates to the car producer (manufacturer) for use in new cars.

Let's look at that process in more detail. From the disassembler, the metal plates go to the materials processor. The disassembler updates the information in the online database about the weight and the plates and the so the materials processor can access it. The paints and coatings are removed and retained. The bare metal parts are combined with others of exactly the same metal, but from various sources. All of this metal is reprocessed together by melting it down to a liquid state. It is then measured out by weight and made into portions, let's say in this case, sheets. The materials processor sends the same amount of metal that was identified by the disassembler as owned by a particular producer to the new-parts supplier to be made into new parts for that producer.

The new-parts supplier makes new metal car bodies for the producer out of the materials received from the materials processor. The new-parts supplier continues to track the weight of the metal and provides that information to the producer along with the new parts. The paint and coatings are also tracked and recombined with captured solvents from the paint rooms in the processor's factory.

Volume is tracked in cases where the used products have some fluids or gases in them that can be reprocessed and reused. Refrigerators have freon gas, for example. The disassembler suctions the gas from the used refrigerator and collects it for the producer.

The *exact same* materials might not be returned to the producer since reprocessing involves batches of materials from various sources and producers. If a particular material is proprietary, it will have to be handled separately.

Because the producer owns the materials, the new-parts supplier bases the cost of the new parts on production, operating costs, and profit margin, and excludes the cost of materials.

Selling Materials

What if the producer wants to sell off used parts (and therefore, the materials in them) and buy new parts or materials for a new product? Here's how it would be handled: the products continue to belong to the producer while they are in the possession of the consumer, but when they reach the distributor/collector, are sent directly to the disassembler. The disassembler then takes the products apart and sells the parts to the used-parts broker. The disassembler sends most of the proceeds to the producer. (We'll discuss income and expenses for all roles in the Revenue Model section.) At the used-parts broker, the parts go through triage (including repair) and are made available for sale, or they are sent to the materials processor. The method for reprocessing the parts is conveyed from the disassembler to the used-parts broker and then to the materials processor or the next buyer as a normal and necessary part of the deal. In the next section, Revenue Model, we refer to used parts that the producer sells as 'disowned'.

Let's look at the *s4ms** for this situation (following).

---

* System for material sustainability

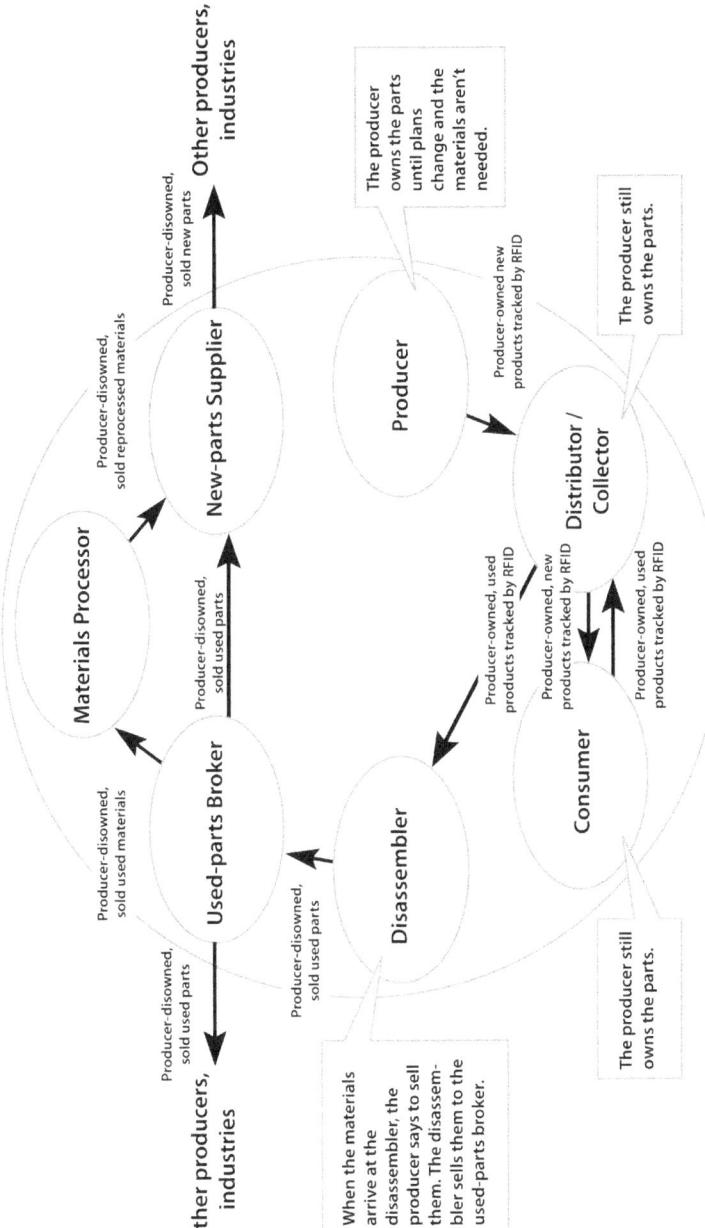

Figure 16: Producer-disowned materials, sold after disassembly.

Purchasing Materials

What about purchasing materials? Since the producer doesn't actually make parts out of elemental materials, he/she purchases materials in the form of parts. In fact, there is usually a succession of parts suppliers that take simple components and assemble them into more complex ones. But for now, let's keep it basic. In this case, the producer buys new parts from new-parts suppliers and used parts from the used-parts broker. Producers may be able to obtain used parts from the disassembler more cheaply than from the used-parts broker. Perhaps the disassembler and the used-parts broker could be merged into one entity. But for the sake of simplicity, we'll say that the disassembler doesn't market used parts. Figure 17 (following) shows how purchased materials come into the possession of the producer.

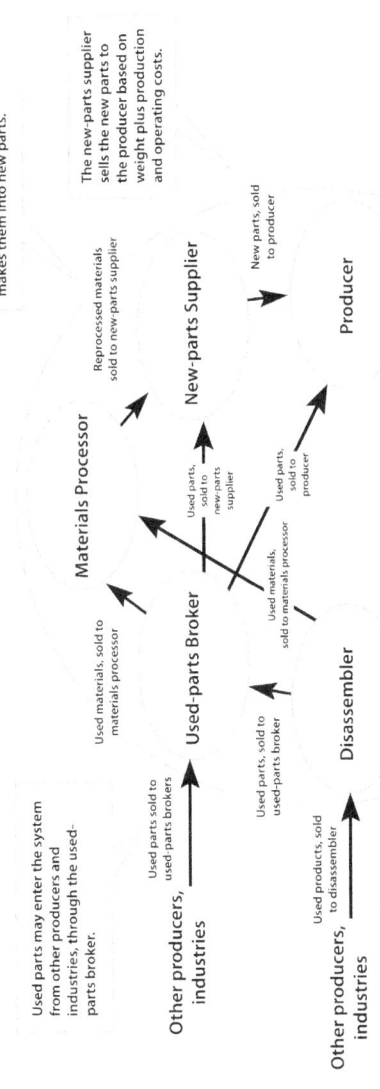

The materials processor sells materials to the new-parts supplier, who makes them into new parts.

The new-parts supplier sells the new parts to the producer based on weight plus production and operating costs.

Reprocessed materials sold to new-parts supplier

Materials Processor

New-parts Supplier

New parts, sold to producer

Producer

Used parts, sold to new-parts supplier

Used parts, sold to producer

Used materials, sold to materials processor

Used materials, sold to materials processor

Used-parts Broker

Disassembler

Used parts may enter the system from other producers and industries, through the used-parts broker.

Used parts sold to used-parts brokers

Other producers, industries

Used parts, sold to used-parts broker

Used products, sold to disassembler

Other producers, industries

Used products may enter the system from other producers and industries through the disassembler. The used products are broken down into used parts by the disassembler.

Figure 17: Purchased materials.

71

The price of used parts is based on the market value of the materials measured in weight or volume and the production and operating costs (which may be devalued after passing through several hands).

If a producer needs more of some item, he/she can check to see if the part is available used from a used-parts broker, but if not, he/she can buy it from or have it made by a new-parts supplier.

Chemical Leasing

In the old linear system of production and consumption, chemical companies or their suppliers sold chemicals to producers, who then became responsible for their use and disposal. Once again, the value of the materials was lost in disposal. Some chemical companies, notably SAFECHEM Europe, a subsidiary of Dow Chemical, have developed the concept of *chemical leasing*. The United Nations Industrial Development Organization (UNIDO) is encouraging these efforts and setting up pilot projects around the world in partnership with various industries[31].

The concept of chemical leasing is service- rather than sales-oriented. In this business model, the producer pays for the benefits obtained from the *use* of chemicals, not for the chemicals themselves. The supplier is no longer concerned with procuring and selling quantities of chemicals for a one-way trip from source to the drain or the air. Now the supplier provides a service in which chemicals are filtered and reused many times.

Chemical leasing is feasible for a wide array of industries and purposes, such as:
- Cooling and heating operations
- Cleaning and purification

- Greasing or degreasing of parts
- Catalyst use
- Powder coating
- Lubricating
- Biocides
- Tin plating
- Caps isolation
- Phosphating[32]

Naturally, chemical leasing business models have the greatest success when they have good recycling rates. A great example of chemical leasing is the SAFE-TAINER™[33] system from SAFECHEM Europe, a subsidiary of Dow Chemical. This system is a closed system for handling chlorinated solvents. Chlorinated solvents are necessary for cleaning metal and are preferred because they are powerful degreasers and are nonflammable. The system combines these advantages with an environmentally friendly closed-loop product cycle that helps conserve resources. With the SAFE-TAINER™ system, producers benefit from a safe, sustainable use of a variety of solvents and methylene chlorides.

A variety of attachnents allow producers to connect the SAFE-TAINER™ system to all types of degreasing machines. The SAFE-TAINER™ system includes two containers. One is for the supply of fresh solvent and the other for the collection of used solvent. The container for fresh solvent is connected to the cleaning equipment in one step. Used solvent, *including vapors*, is returned to the used solvent container. The used solvent is collected and returned to SAFECHEM for recycling. With this system, virtually all risks of emissions and spills during transfer and use of the solvent are eliminated. SAFECHEM reports that through additional services and advice to producers, a solvent lifespan can be substantially prolonged and solvent consumption can be reduced by up to 80%[34].

Even better than that, Dow Chemical is taking recovery to a new level by offering to help producers recover a wide range of solvents *whether they are Dow products or not*. Producers can either reuse these recovered solvents themselves or give them to Dow. Dow's affiliate, Dow Haltermann Custom Processing, has expertise in recovering and recycling solvents from waste or byproducts. They handle and transport highly volatile and hazardous solvents, process and sell them and safely dispose of residue. This service provides a professional method for recovery or disposal of solvents for producers.

Dow Haltermann Custom Processing recovers and reprocesses these kinds of solvents:
- Acetone
- Dimethyl Formamide (DMF)
- Industrial Methylated Spirits (IMS 96)
- Ethanol 96
- Isopropanol (IPA 99% or 87%)
- Monoethylene Glycol (MEG)
- Methyl Ethyl Ketone (MEK)
- N-Methyl-2-Pyrrolidone (NMP)
- Tetrahydrofuran (THF)
- Toluene[35]

Product designers/manufacturing engineers, disassemblers and materials processors should take advantage of new developments in chemical leasing for disassembly challenges related to paint, coatings, grease and other chemicals that would otherwise be impossible to solve.

Chemical reuse should be extended to all chemicals used in manufacturing. Producers dispose of hundreds of materials that could potentially be recovered and reused but perhaps currently it is easier and less expensive to dispose of them. Producers who use any of 650 toxic chemicals are required to report to the Environmental Protection Agency (EPA) and state governments annually as to whether these chemicals were used, manufactured, treated, transported to off-site locations, or released into the environment. The EPA compiles these data in an online, publicly-accessible national database called the Toxic Release Inventory (TRI). These chemicals include such things as aluminum, ammonia, boron, chlorine, copper, fluorine, formaldehyde, freon, methanol, nitroglycerin, salt, toluene, and zinc.

It may be interesting for both producers and consumers to see what materials and chemicals producers discard. Consumers can access the database and enter their zip code to see what materials and chemicals producers are disposing of in their area. Visit and explore the TRI website[36].

Summary – Materials Management

In systems for material sustainability, materials are owned and tracked from the time they leave the producer to their return, either as used parts from the disassembler or new parts from the new-parts supplier via the materials processor. RFID aids in tracking products while they are in the hands of the distributor/collector or the consumer and up to the time they arrive at the disassembler. After that point, interested parties track parts by part number or they track materials by weight or volume in an online database.

Materials are not allowed to escape the *s4ms**– not even the fumes.

Research and Development Directions for Materials Management

Here are some challenges for research and development for materials management:

- Design a logistics system that minimizes costs for transportation among the roles in the system for material sustainability.

- Design a system and online database to manage materials that can be shared by all the roles.

- Create a computer model to simulate the movement of materials throughout the *s4ms**.

---

* System for material sustainability
* System for material sustainability

Revenue Model

The revenue model supporting the *s4ms** is based on *leasing a property rather than on selling a product*. Other authors have written about the innovative practice of selling a service, even when it involves a tangible product of some kind, rather than selling the product itself. Such companies as Carrier, Electrolux, and Schindler already provide air conditioning, refrigeration, and even elevators as *services*, rather than as products that are sold[37]. Yet a better concept may be that of leasing a property, as this clarifies ownership of the materials involved.

As in any lease agreement, producers provide repair or replacement services through their distributors/collectors. They also handle swaps and upgrades for consumers. If consumers want to return, swap, or upgrade products, the used products can be leased out to second-tier consumers who perhaps can't afford or don't care to have the latest model. Let's call this the 'extended life' of a product. Producers can continue to make money on used products at a lower payment scale. This practice also permits consumers who want the latest model of the product to have it earlier, since another consumer assumes the payments on the older model. Those who want the latest model may pay a premium. Producers may also charge a penalty to consumers for leaving their product line, much like cell phone providers do now.

The benefit to producers is the retention of materials and the constant income stream. The benefit to consumers is use of a high-quality, reliable product. Second-tier markets create an incentive for producers to make high-quality, reliable products.

---

* System for material sustainability

Do systems for material sustainability discourage innovation? Not at all. The market is as open as ever, and consumers can switch to other products. Other producers may pay a bonus to new consumers if they switch, thereby offsetting the penalty assessed by the previous producer.

How does each of the roles in the *s4ms** make money? Let's take a look at income and material flows. Figure 18 (following) shows income flows as dotted-line arrows and material flows as solid-line arrows between roles.

---

* System for material sustainability

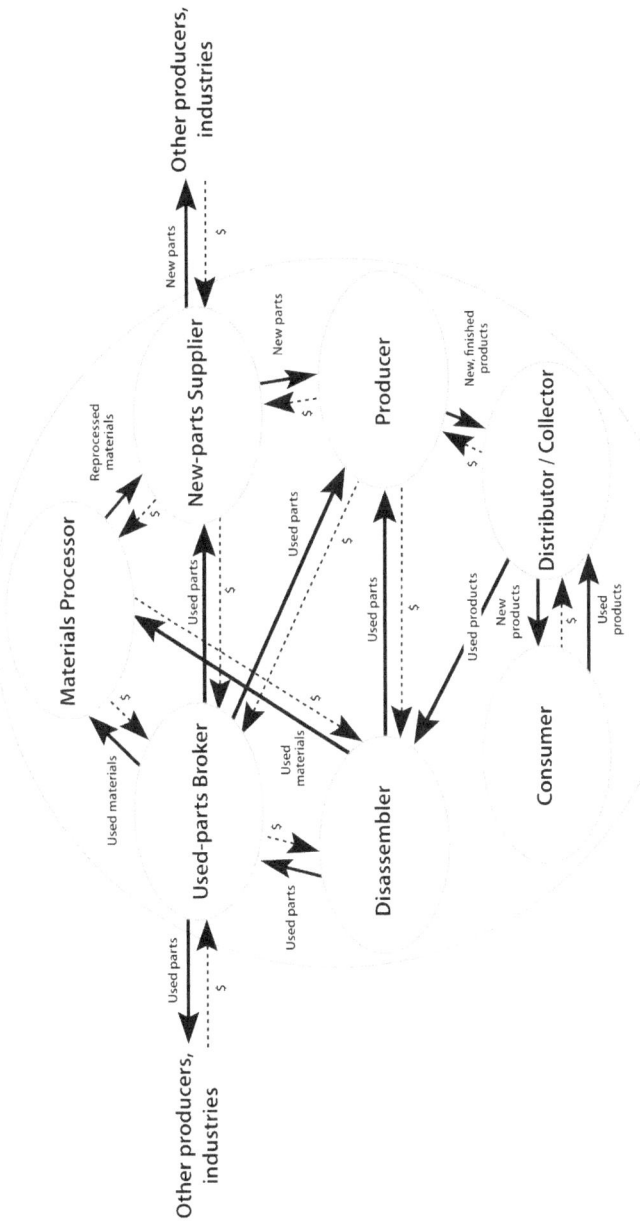

Figure 18: Income flows (dotted arrows) in the system for material sustainability (s4ms).

This gives us a broad overview. However, we need to look at each role individually to understand income and expenses. Let's start with the consumer. The vast majority of consumers rely on jobs for their income. A healthy consumer base requires plenty of jobs that pay a middle-class income. Of course, that opens up bigger issues, outside the scope of this book. But depending on the product, there may be thousands, tens of thousands, or hundreds of thousands of consumers, all paying a monthly fee for the use of the product. *The consumer is the source of the income for the whole system.*

Next, let's jump over to the producer. The producer's income stream comes from the consumers through the distributor/collector. The producer's expenses include production costs, operating costs, purchase of materials and new and used parts, and repair of parts returning from the disassembler. Equipment, computer systems, and RFID technology are substantial expenses. There are of course, many other normal business expenses but we won't go into all of them.

The distributor/collector applies a surcharge or fee to consumers. This fee supports marketing, inventory and tracking, collection, repair, consumer account management, and transportation of used products to the disassembler. The distributor/collector may or may not serve multiple producers.

The income for the new roles of disassembler and materials processor comes mainly from fees charged to producers. Although income for these roles is predicated on the value of materials, recall that earlier in this book we saw that Xerox saved several hundred million dollars each year because they used recycled materials rather than raw materials to produce new machines. Also, for the disassembler and the materials processor, economies of scale can be achieved if they serve multiple producers and if disassembly and materials reprocessing processes and equipment can be adaptable for multiple products.

Although the disassembler's income comes mostly from fees charged to the producer, some comes from the sale of parts to the used-parts broker and the sale of materials to the materials processor. Here again, equipment, computer systems, and RFID technology are substantial expenses.

At this point, we have to make a distinction between materials that the producer wants to *continue to own* versus those that he/she *decided to sell*. The disassembler sends owned used parts back to the producer, who pays the disassembler a fee for the service. The disassembler sells parts that the producer no longer wants, the *disowned* parts, to the used-parts broker or materials processor, depending on the condition of the parts. Income from the sales of disowned parts is split between the disassembler and the producer.

The used-parts broker's income comes from the sale of used parts and materials. Expenses come from the purchase of used parts from the disassembler and other normal business expenses. The used-parts broker may repair parts if needed or relegate that task to the producer, depending on demand. One thing the used-parts broker doesn't have to worry about is tracking producer-owned parts and materials. The former owners of these used parts have released their interest in them.

Materials come to the materials processor from the disassembler as producer-owned or -disowned. The materials processor buys disowned materials from disassemblers and used-parts brokers, reprocesses the materials and sells them to new-part suppliers. The sale of disowned materials includes the value of the materials and the reprocessing fee. If the materials are owned, the materials processor charges a fee to the producer for reprocessing his/her materials before transferring them to the new-parts supplier to be made into new parts.

Materials come to the new parts suppliers as producer-owned or -disowned. The new-parts supplier buys disowned materials from the materials processor and makes new parts for sale to various producers. Since the materials are disowned, the sale of the parts includes the value of the materials. If the materials are owned, the new-parts supplier charges a fee to the producer to make parts out of his/her materials but the sale doesn't include the value of the materials themselves.

And just to emphasize it again, when the parts are producer-owned, *none* of the roles can charge the producer for the value of the materials. They can only charge a fee for services.

Now we have looked at each role in the system. As mentioned earlier, almost every role can serve multiple producers and industries and thereby make a good income. But to do this, each role has to think in terms of offering a wide range of services. Equipment used by disassemblers, materials processors, new-parts suppliers should be highly flexible and programmable and usable for a wide variety of tasks. Even producers could cut costs by making use of manufacturing services provided by factories that serve multiple producers and industries.

Designers of disassembly processes and equipment must work together very closely with producers and new-parts suppliers to identify requirements for disassembly methods. In time, methods can be developed that are capable of handling a wide variety of products in a cost-effective way.

Another way to minimize expenses for the service providers is to have all of their services under one roof. The disassembler, materials processor, new-parts supplier and possibly the used-parts broker could all be in one location – even one company – and save on costs related to computer systems, equipment, operations and transportation. If assembly and disassembly could be combined in one factory, perhaps equipment used for assembly of new parts and products could also be adapted for disassembly.

Summary – Revenue Model

In an *s4ms**, producers no longer *sell a product*, they *lease a property*. Income from leased products is extended because used products that are in good condition are leased to second-tier consumers.

Many of the roles in the *s4ms* continue to make money as they always have. But the new roles, especially the disassembler and the materials processor, earn income mainly from fees they charge to producers. Because of this, they need to serve as many producers as possible with flexible, programmable equipment and processes that can handle a wide variety of products, parts, and materials.

Research and Development Directions for Revenue Model

Here is a key challenge for research and development of a revenue model:

- Create a computer model of the income and expenses for all roles. Optimize it to enhance the income of all roles.

---

* System for material sustainability

Next Steps

How can a business or industry actually get started in building an *s4ms**? Let's look at it this way: some of the changes are easier to tackle than others. In terms of changing a revenue model, the transition from the notion of selling a product to leasing a property is one of the easier challenges. There are plenty of examples that can be adapted.

RFID is already being deployed in numerous applications and different types of RFID tags with varying levels of capabilities are already available. The participating roles in an *s4ms* work together to specify the kinds of information they need to share and the kind of computer systems that support their needs.

Remanufacturing is also not new, and many producers already practice it. Only now remanufacturing must merge into and become identical with manufacturing.

The major challenges for an *s4ms* are in the areas of materials engineering, product design, assembly and disassembly, and materials reprocessing. All of these must be considered, coordinated, and optimized together. None of these functions can operate alone in a system for material sustainability.

The amount of work needed to make the transition to an *s4ms* is too great for any one producer. But once the system is set up, the advantages will be great.

This is where the Institute for Material Sustainability can help. As we mentioned earlier, the purpose of the institute is to provide services to industries that will help them make

---

* System for material sustainability

the transition to systems for material sustainability. The institute will serve as a focal point for experts who collaborate on the development of systems for material sustainability and then serve as consultants to industries. Specialists are needed in each of the areas mentioned above: materials engineering, product design, assembly and disassembly, materials reprocessing and system optimization. The institute will identify projects and challenge research and development specialists in both private and public organizations to find solutions. The institute will also aid in coordinating funding from foundations or other grantors for the completion of projects. Here are the goals in each area:

- **Materials engineering:** Develop perpetually reusable materials to replace 80% of mongrelized materials currently in use in the next 10 years.

  Examples of projects:
  - Rigid plastics – used for TVs, stereos, laptops, office furniture
  - Paint – used on metal, glass, plastic
  - Non-stick coatings – used in cookware
  - Rubber – used in tires
  - Paper – used for business, publishing and personal needs – eliminate the need for new supplies of wood or replace with plastic
  - Ink – used on paper – create a type of ink that can be removed and recovered
  - Soft plastics – used for shopping bags, toys, sheet plastic, shrink-wrap
  - Steel – used in construction, vehicles, cookware
  - Cement – used in construction, streets and sidewalks

- **Product Design:** Determine how to use perpetually recyclable materials to design new products that replace current products. Select 50 products as pilot projects.

- **Assembly and Disassembly:** Design assembly and disassembly processes and equipment, write disassembly plans and suggest reuses for parts and materials. Design disassembly and reprocessing plants.

- **Materials Reprocessing:** Design materials reprocessing processes and equipment, write plans for reprocessing materials.

- **System Optimization:** Elaborate fully the entire *s4ms** for multiple industries. Create a computer model that encompasses all the roles in the *s4ms* including materials management and revenue model. Set up RFID and database system for an *s4ms*. Set up a pilot disassembly plant and a pilot materials reprocessing plant with equipment, robotics, computer systems and personnel. Conduct full system tests.

If we get the support we need, we will conduct a full system test to try out one complete cycle within an *s4ms*. The institute's experts will select a product or products for the test. The test will be based on the computer model used to optimize the system. The test may involve thousands of people playing the roles and performing the processes of an *s4ms*. Real materials and monies will be used. All activities will be carefully observed and analyzed. The results will be described in an evaluation report, which will then be used

---

* System for material sustainability

to improve the system. Following evaluations, there may be additional iterative full system tests to resolve issues.

As the knowledge base develops and reaches maturity, the institute's experts will provide consulting services to producers and supply chain members to help them make the transition to a system for material sustainability.

Let's Start the Dialog

Please go to the Institute for Material Sustainability website (www.s4ms.org) and contribute your ideas and expertise in any of the areas mentioned in this book. We are in the process of establishing the Institute for Material Sustainability as a non-profit organization. We need board members, experts, specialists, research facilities, and funding. If you are in a position to offer expertise or funding, or if you know of other people or organizations who may be interested in creating a way to recycle everything, please contact the institute at the website.

Websites:

www.s4ms.org
(the Institute's website)

www.rebk.org
(the book's website)

About this Book

This book began as a paper I wrote for my master's degree in engineering and technology management in 2001. I continued to work on it for several years, collecting information and reading books on the subject of sustainability. I searched through all of the books listed in the reading list (below) and countless articles on the subject of sustainability, and although these authors and editors made invaluable contributions to the growing awareness of the need for sustainable systems, I wasn't able to find a complete, whole system, a closed-loop system for sustainable production and consumption such as the one described in this book.

I feel that we human beings need to change the way we operate. We need a vision of what a new way could look like and how it could work. This book does not begin to elaborate all of the details that will have to be worked out to make it possible to recycle everything. However, I believe that once a thing is imagined, the rest is engineering.

About the Author

Ms. Unruh holds a master's degree in Engineering from Portland State University. She is currently the training manager in the manufacturing engineering department of an international truck manufacturer. She lives in Portland, Oregon with her husband and son.

Notes

Evidence of Declining of Material Availability

List of Materials and Years to Depletion from *New Scientist*[38].

Material	Years to Depletion[39]	Uses	Rate of Recycling
Antimony	13-30	Drugs, flame retardants, paints, ceramics, enamels, a wide variety of alloys, electronics, and rubber	-
Chromium	40-143	Chrome plating, paint	25%
Copper	38-61	Wire, coins, plumbing	31%
Gold	36-45	Jewelry, dental	43%
Indium	4-13	Flat-panel TV and computer screens	-
Lead	8-42	Batteries	72%
Nickel	57-90	Batteries, turbine blades	35%
Platinum	42-360	Jewelry, catalysts, fuel cells for cars	-
Silver	9-29	Jewelry, catalytic converters	16%
Tantalum	20-116	Electronic components, cell phones, camera lenses	20%
Tin	17-40	Cans and solder	26%
Uranium	19-59	Weapons, power stations	-
Zinc	34-46	Galvanizing metals	26%

Table 2: List of materials and years to depletion.

Critical Minerals and the U.S. Economy

Minerals, Critical Minerals, and the U.S. Economy (2008) was written by the Committee on Critical Mineral Impacts of the U.S. Economy, Committee on Earth Resources, National Research Council and is available from the National Academies Press[40]. The book is the result of a study whose purpose was to determine which minerals could be considered critical to the nation's economy and what research may be needed to help mitigate disruptive fluctuations in the supply of these minerals to industry. According to the book:

> Minerals are part of virtually every product we use. Common examples include copper used in electrical wiring and titanium used to make airplane frames and paint pigments. The information age has ushered in a number of new mineral uses in a number of products including cell phones (e.g., tantalum) and liquid crystal displays (e.g., indium). For some minerals, such as the platinum group metals used to make catalytic converters in cars, there is no substitute. If the supply of any given mineral were to become restricted, consumers and sectors of the U.S. economy could be significantly affected. Risks to minerals supplies can include a sudden increase in demand or the possibility that natural ores can be exhausted or become too difficult to extract. Minerals are more vulnerable to supply restrictions if they come from a limited number of mines, mining companies, or nations. Baseline information on minerals is currently collected at the federal level, but

no established methodology has existed to identify potentially critical minerals. This book develops such a methodology and suggests an enhanced federal initiative to collect and analyze the additional data needed to support this type of tool[41].

The book identified eleven minerals that were deemed critical to industry and assessed the impact on the economy if the supply of any of these minerals were restricted and the potential risk that supplies actually could be restricted. The criticality matrix diagram (below) shows the two main factors that determine the scoring of a mineral's criticality: the *impact* of supply restriction as determined by its importance in use and industry's ability to use substitutes (Y axis), and the *supply risk* which involves potential factors affecting the availability of the mineral (technological, political, economic, etc.) (X axis).

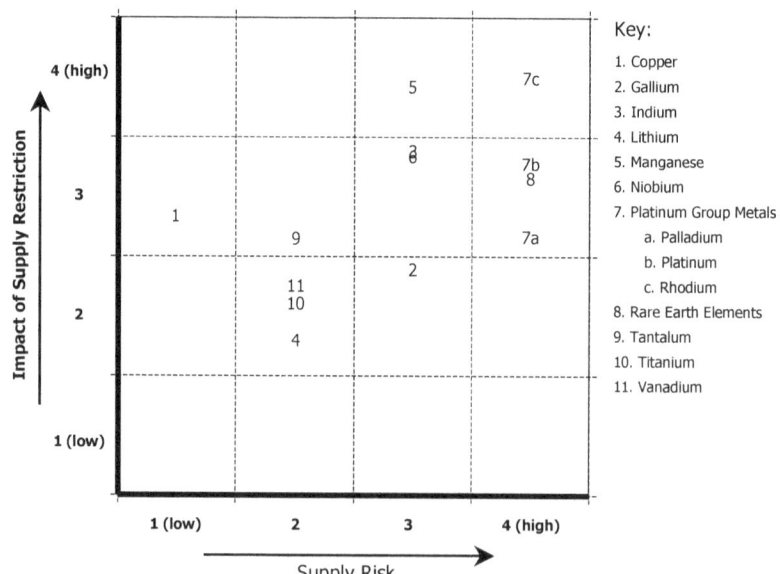

Chart 1: Impact and Risk of Supply Restriction of 11 Essential Minerals, adapted from Minerals, Critical Minerals, and the U.S. Economy[42].

The preceding chart shows that in the case of several of these minerals (or materials), the impact of a shortage on the economy would be extremely high and the risk that supplies could be in danger are also high – the prime example being Rhodium (**7c** in the upper right corner of the chart). For more information on the study and its conclusions, you can read the entire book on the web at www.nap.edu/catalog/php?record_id=12034.

Basic Materials Prices

For the past 10 years, the prices for basic materials were on an upward trend – until they were knocked down by the recent economic crisis. But now they are on the rise again. This is an excellent reason for adopting an *s4ms**.

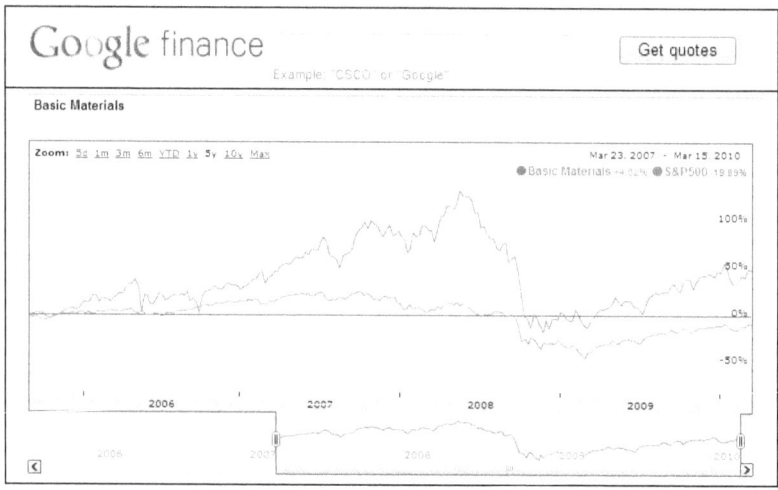

Chart 2: Basic materials prices (in percent)[43]. Accessed 15 March 2010.

---

* System for material sustainability

U.S. Dependence on Foreign Sources of Vital Materials

To what extent is the U.S. dependent on foreign sources for vital materials? The United States Geological Survey (USGS) collects this data and publishes an annual report called the Mineral Commodities Summary. The following chart is from the most recent edition[44].

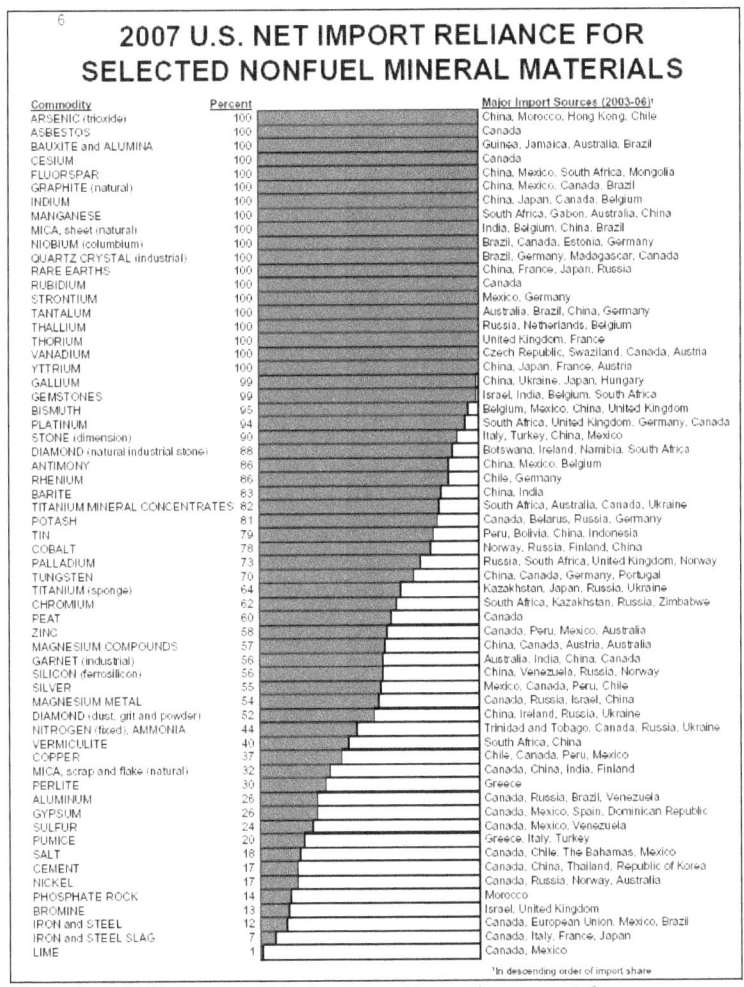

Chart 3: USGS data on U.S. import reliance for materials.

This chart (preceding) shows U.S. dependence on foreign sources for quite a number of commodities. If relations with any of these countries turn sour, supplies may be threatened. USGS information contains statistics and information on the worldwide supply of, demand for, and flow of minerals and materials essential to the U.S. economy, the national security, and protection of the environment.

Vital Materials and National Security

And finally, one would think that if there were something that could have a large impact on the economy and possibly even national security, that the U.S. Department of Defense would be involved. And in fact, they are. A book on this topic entitled, *Managing Materials for a Twenty-first Century Military (2008)*[45], was written by Committee on Assessing the Need for a Defense Stockpile, National Research Council and is available from the National Academies Press.

According to the book, the U.S. government has maintained a supply of critical strategic materials since 1939 in the National Defense Stockpile (NDS). The NDS is required by law to store strategic and critical materials in the interest of national defense to reduce dependence on foreign sources in a national emergency. Federal law mandates the agency to hold materials for all essential civilian and military uses in times of emergency. The NDS is keenly aware that the needs of the military have changed. Many new kinds of technology require different materials from those stockpiled in the past. The NDS is focusing on these kinds of strategic materials:

- Stealth technology materials
- Electronic and photonic materials for high-speed communications
- Sensor and actuator materials
- High-energy-density materials
- Materials that improve propulsion technology
- Lightweight materials that provide functionality equivalent to that of their heavier analogs
- Materials that enhance protection and survivability.[46]

The need for such a stockpile is exacerbated by global competition for materials. Developing countries are growing at an accelerating rate. Economic data from the International Monetary Fund (IMF) for the years 1950 through 2004 show that in a little more than 50 years, the global economy grew from $7.1 trillion to $56 trillion in constant (inflation adjusted) U.S. dollars. The book predicts that by 2040 the combined economies of Brazil, Russia, India and China (referred to as the BRIC economies) will be larger than the combined economies of France, Germany, Italy, Japan, the United Kingdom, and the United States. The increasing demand has pushed mineral prices up to new highs over the last 10 years.

The NDS is adapting to the increasing global demand for minerals, the dramatic changes in source locations, volatile markets and prices, corporate mergers in the global mineral industry, and increased vulnerabilities in the mineral supply chain.

As of May 2007, the Defense National Stockpile Center (DNSC) stored 21 materials at locations throughout the United States. These included platinum, used for chemical catalyst applications, including catalytic converters; germanium, used for detectors, fiber-optic systems, and infrared optics; and ferrochrome, a metal additive used in stainless steel and other specialized alloys.

Stockpiles are finite; they can't be a permanent or long-term solution to shortages. At best, the NDS can hope to serve us in a short-term emergency. But the fact that the U.S. Department of Defense deems it necessary to have a stockpile of materials shows that the issue of materials availability is serious.

Concepts Related to Sustainability

Many concepts have gained popularity in the past few decades and they have all made important contributions. But none have envisioned a way out of the linear production-consumption model. Mostly they are based on the concept of eco-efficiency, which applies regulatory controls to reduce energy use, material consumption, and impacts upon the environment. Eco-efficiency doesn't *question* the linear model; it simply slows down the process. The end result, a Malthusian decline, is unavoidable. It would simply be farther into the future. Let's look at some of the popular concepts briefly.

Design for Environment is a systemic approach to decision support for designers analyzing all potential environmental implications of a product or process, its energy and materials used, manufacture and packaging, transportation, consumer use, reuse, recycling, or disposal. It includes the traditional design issues of cost, quality, and manufacturing processes in its decision system. This approach integrates life-cycle analysis, concurrent engineering, and design for X. Using a matrix of question sets, designers score each activity against environmental concerns. The matrix provides the design team a way to see the whole project and its parts. Two AT&T executives, Braden Allenby and Thomas Graedel, describe this system in their book, *Industrial Ecology*[47].

Dynamic Input-Output Modeling[48] enables business and policy makers to grasp the implications of business, economic and environmental change over time. Based on the Nobel-Prize-awarded work of Wassily Leontief[49], Faye Duchin[50] created a means of analyzing the total impacts of alternative scenarios of industrial change. Dynamic input-output models are used to weigh complex trade-offs and develop a set of best-case scenarios.

Industrial Ecology is an interdisciplinary field that focuses on the sustainable combination of environment, economy and technology. The central idea is that industrial systems should incorporate principles observed in natural ecosystems.

Industrial Metabolism traces material and energy flows from initial extraction of resources through industrial and consumer systems to the final disposal of wastes.

Dr. Robert Ayres[51] first developed this form of analysis in the 1970s, and it has become an important foundation of industrial ecology.

Life Cycle Analysis identifies and quantifies energy and materials flows in an industry from cradle to grave, including emissions and wastes, much like industrial metabolism. Life cycle analysis goes further, however, and assesses the consequences for the environment and identifies areas for improvement that can reduce environmental burdens. Volvo's Environmental Priority Strategies system[52] is a sophisticated application of life cycle analysis that assists product designers in determining which choices will have the least negative impact on the environment.

Life cycle analysis is the basis for ISO 14000 certification.

The Natural Step uses life cycle analysis and the 'system conditions' to assist businesses and communities in becoming more sustainable. The system conditions are four conditions that must be met to be sustainable. In each, nature is not subject to an increase in: 1) concentrations of substances extracted from the earth's crust, 2) concentrations of substances produced by society, 3) degradation by physical means and 4) people are not subject to conditions that systematically undermine their capacity to meet their needs. The Natural Step also promotes the concept of the triple bottom line including economic, ecological and social bottom lines. The triple bottom line recognizes that there are wider-scope costs and responsibilities beyond profitability.

Reading List

TITLE	AUTHORS
Agenda 21; the Earth Summit Strategy to Save Our Planet	Sitarz, editor
Cradle to Cradle; Remaking the Way We Make Things	McDonough and Braungart
Creating a Sustainable Future; Learning to Change Your Life, the Community and the World	Francis
Discovering Industrial Ecology; and Executive Briefing and Sourcebook	Lowe, Warren and Moran
Eco-Economy; Building an Economy for the Earth	Brown
Ecology of Commerce; A Declaration of Sustainability, The	Hawken
Ecology of Industry; Sectors and Linkages	Richards and Pearson, editors
Economics of Natural Resources and the Environment	Pearce and Turner
Factor Four; Doubling Wealth, Halving Resource Use	Weizsacker, Lovins and Lovins
Industrial Ecology (2nd Edition)	Graedel and Allenby
Industrial Ecology and Global Change	Socolow, Andres, Berkhout and Thomas, editors
Industrial Ecology; Towards Closing the Materials Cycle	Ayers and Ayers
Last Oasis; Facing Water Scarcity	Postel

Continued, next page

TITLE	AUTHORS
Limits to Growth: The 30-Year Update	Meadows, Randers and Meadows
Managing Materials for a Twenty-first Century Military	National Research Council
Minerals, Critical Minerals, and the U.S. Economy	National Research Council
Natural Capitalism; Creating the Next Industrial Revolution	Hawken, Lovins and Lovins
Natural Step for Business; Wealth, Ecology and the Evolutionary Corporation, The	Nattrass and Altomare
One With Nineveh: Politics, Consumption, and the Human Future	Ehrlich and Ehrlich
Our Ecological Footprint; Reducing Human Impact on the Earth	Wackernagel and Rees
Pillar of Sand; Can the Irrigation Miracle Last?	Postel
Resource Wars; The New Landscape of Global Conflict	Klare
State of the World 2004: Special Focus: The Consumer Society	Worldwatch Institute
State of the World 2008: Innovations for a Sustainable Economy	Worldwatch Institute
Tapped Out; the Coming World Crisis in Water and What We Can Do About It	Simon

Table 3: Books on the subject of sustainability.

References

[1] _____, "Metals, the New," *AZoM™ - The A to Z of Materials* (no date), http://www.azom.com/News.asp? NewsID=168. Accessed 2 November 2008.

[2] David Cohen, "Earth's Natural Wealth, An Audit," *New Scientist: Environment* (23 May 2007), http://environment. newscientist.com/ article/mg19426051.200. Accessed 2 November 2008. Sources of data: Armin Reller, University of Augsberg Germany, Tom Graedel, Yale University and the U.S. Geological Survey. The data is summarized in a table in the Notes section of this book.

[3] Neil V. Gayle and Dave Anderson, "Materials Supply Chain Management Efforts at ISMT," *Semiconductor International* (1 March 2003), http://www.semiconductor. net/index.asp? layout=article Print&articleID=CA280935. Accessed 2 November 2008.

[4] _____, "Continued Depletion of Lead Signifies a Cause for Concern," *Connecting Industry* (16 April 2008), http://www.connectingindustry.com/story.asp?storycode =184880. Accessed 2 November 2008.

[5] Robert Burns, "High Cost of Nitrogen Calls for New Farming Strategies," *AgNews* (16 April 2007), http://newagnews.tamu.edu/dailynews/stories/SOIL/Apr16 07a.htm. Accessed 2 November 2008.

[6] Wikipedia contributors, "Club of Rome," *Wikipedia, The Free Encyclopedia* (14 November 2008). http://en.wikipedia.org/w/index.php?title=Club_of_Rome& oldid=251829368. Accessed 16 November 2008.

[7] _____, "About Us: History, the Story of the Club of Rome," *The Club of Rome Website* (2008). http://www. cluboframe.org/. Accessed 16 November 2008.

---

[8] Donella H. Meadows, *Limits to Growth* (Signet, 1972). See also Donella H. Meadows, Jorgen Randers, Dennis L. Meadows, and William W. Behrens, *The Limits To Growth: A Report For The Club Of Rome's Project On The Predicament Of Mankind, Second Edition* (Universe, 1974).

[9] Donella H. Meadows, Jorgen Randers, Dennis L. Meadows, *Limits to Growth, the Thirty-Year Update* (Chelsea Green, 2004).

[10] Within the last two months, a worldwide economic crisis has begun and the price of materials is falling through the floor. If the extraction business becomes unprofitable, this is an even greater reason to adopt Recycle Everything because it will insulate manufacturing from a diminished availability of materials. This is explained in the next several pages. Please continue reading.

[11] McDonough Braungart Design Chemistry website, http://www.mbdc.com/.

[12] Robert U. Ayres was Professor of Engineering and Public Policy at Carnegie-Mellon University from 1979 to 1992, and then moved to the European business school INSEAD, in France, where he is now Emeritus Professor of Environment and Management. He is a brother of WORLD WATCH editor Ed Ayres. See the book list in the Notes section.

[13] It remains to be seen whether current efforts to optimize the linear system, such as those related to the supply chain, lean manufacturing, pull versus push, six sigma and others are adaptable to system for material sustainabilitys or they are better served by new concepts.

[14] Land Institute website, http://www.landinstitute.org/. Accessed 2 November 2008.

[15] Lisa M. Hamilton, "Cultivating Long-term Relationships", *Rodale Institute* (13 September 2004), http://newfarm.rodaleinstitute.org/features/0904/perennialwheat/index.shtml. Accessed 2 November 2008.

[16] Tim Steury, "Full Circle: Perennial Wheat Could Fulfill a Tradition and Transform a Landscape," *Washington State Magazine Online* (no date), http://washington-state-magazine.wsu.edu/stories/04-summer/wheat/index.html. Accessed 2 November 2008.

[17] Land Institute website, http://www.landinstitute.org/vnews/display.v/ART/2007/03/15/45facffb6ccd6. Accessed 2 November 2008.

[18] Trudy A. Dickneider Ph.D., "Petretec — DuPont's Technology for Polyester Regeneration," *Green Module for Industrial Chemistry* (no date), http://academic.scranton.edu /faculty/CANNM1/industrialchemistry/industrialchemistrymodule.html. Accessed 2 November 2008.

[19] P.M. Morse, "PET Producers Face Rough Transition Market", *Chemical and Engineering News* (22 July 1998), pp 33-35.

[20] DuPont Corporation website, http://www.technologybank.dupont.com/t2h/page/homepage (enter 'petretec' in the search box). Accessed 31 October 2008.

[21] _____, "Waste: LIFEnews Features 2007", *Europa* (no date) http://ec.europa.eu/environment/life/themes/waste/features2007/elves.htm. Accessed 2 November 2008. Also, see the Recieder website, http://www.recieder.com/.

[22] Shaw Floor website, http://www.shawfloors.com/about-shaw/carpet-recycling. Accessed 2 November 2008.

[23] Reva Revis (contact person), "InterfaceFLOR First to Produce Post-Consumer Nylon 6,6 Carpet for Commercial

---

Market," (press release) (11 June 2007), http://www. interfaceflooring.com/pdf/ 06_11_07_Interface_ FLOR_Recycled_Fiber_Press_Release.pdf. Accessed 2 November 2008.

[24] Peter M. Senge and Goran Carstedt, "Innovating our way to the next industrial revolution," *MIT Sloan Management Review* (Winter, 2001), http://sloanreview.mit.edu/ smr/issue/2001/winter/2/. Accessed 2 November 2008.

[25] _____, "Xerox Corporation 1999 Environment, Health and Safety Progress Report," Xerox Corporation (1999), http://www.xerox.com/downloads/usa/en/1/1999ehsprog.pd f. Accessed 2 November 2008.

[26] _____, "2007 Report on Global Citizenship," Xerox Corporation (2007), http://www.xerox.com/ downloads/ usa/en/x/Xerox_Global_Citizenship_Report_2007.pdf. Accessed 2 November 2008.

[27] Ibid.

[28] Ibid.

[29] Active Disassembly website, http://www.activedisassembly.com/. Accessed 2 November 2008.

[30] The alpha-numeric numbers used in this example are for illustration purposes.

[31] United Nations Industrial Development Organization (UNIDO) Chemical Leasing website, http://www chemicalleasing.com/sub/concept.htm. Accessed 2 November 2008.

[32] United Nations Industrial Development Organization (UNIDO) Chemical Leasing website, http://www. chemicalleasing.com/sub/pilot.htm. Accessed 2 November 2008.

[33] Dow Chemical Corporation website, http://www.dow.com/safechem/product/safetain.htm. Accessed 2 November 2008.

[34] Dow Chemical Corporation website, http://www.dow.com/ safechem/pdfs/news_release_final1.pdf. Accessed 2 November 2008.

[35] Dow Chemical website, http://www.dow.com/custproc/capabil/solvrec.htm?filepath=&fromPage =BasicSearch. Accessed 2 November 2008.

[36] Toxic Release Inventory website, http://www.epa.gov/tri/. Accessed 2 November 2008. For specific chemicals, see http://www.epa.gov/triexplorer/. For specific facilities, see http://www.epa.gov/enviro/ html/tris/tris_query.html.

[37] Paul Hawken, Amory Lovins, and L. Hunter Lovins, *Natural Capitalism: Creating the Next Industrial Revolution* (Little, Brown and Company, 2000), 135-138.

[38] David Cohen, "Earth's Natural Wealth, An Audit," *New Scientist: Environment* (23 May 2007), http://environment.newscientist.com/article/mg19426051.200. Accessed 2 November 2008.

[39] The smaller number indicates the anticipated number of years to depletion if world consumption is equal to the U.S. rate and the larger number indicates the anticipated number of years to depletion if world consumption is half of the U.S. rate.

[40] National Academies Press website, http://www.nap.edu/.

[41] Committee on Critical Mineral Impacts of the U.S. Economy, Committee on Earth Resources, National Research Council, *Minerals, Critical Minerals, and the U.S. Economy* (Washington, DC: The National Academies Press, 2008), Description. See http://www.nap.edu/

catalog.php?record_id=12034. Accessed 10 November, 2008.

[42] Ibid, page 111.

[43] Chart, "Basic Materials," Google website. http://finance.google.com/finance?catid=57629812. Accessed 17 November 2008.

[44] United States Geological Survey (USGS) website, http://minerals.usgs.gov/minerals/pubs/mcs/2008/mcs2008. pdf, page 9. Accessed 10 November 2008. For more information, see http://minerals.usgs. gov/minerals/pubs/commodity/.

[45] Committee on Assessing the Need for a Defense Stockpile, National Research Council, *Managing Materials for a Twenty-first Century Military* (Washington, DC: The National Academies Press, 2008). See http://books.nap.edu/catalog.php?record_id=12028.

[46] Ibid, page 70.

[47] Thomas E. Graedel and Braden R. Allenby, *Industrial Ecology* (2nd Edition), Prentice-Hall International Series in Industrial and Systems Engineering, 2008.

[48] See http://en.wikipedia.org/wiki/Input-output_model.

[49] Wassily Leontief earned the Nobel Prize in Economics in 1973 for his work on input-output tables. Input-output tables analyze the process by which inputs from one industry produce outputs for consumption or for inputs for another industry. With the input-output table, one can estimate the change in demand for inputs resulting from a change in production of the final good.

[50] Faye Duchin, Dean of the School of Humanities and Social Sciences, Professor of Economics and prolific author, has created 'what if' tools upon the foundation of

industrial metabolism and structural economics. These dynamic input-output models enable business and policy decision-makers to perceive the broad business, economic, and environmental implications of systemic technical change.

[51] Robert U. Ayres was Professor of Engineering and Public Policy at Carnegie-Mellon University from 1979 to 1992, and then moved to the European business school INSEAD, in France, where he is now Emeritus Professor of Environment and Management. He is a brother of WORLD WATCH editor Ed Ayres. See the book list in the Notes section.

[52] Bengt Steen, "A Systematic Approach to Environmental Priority Strategies in Product Development (EPS). Version 2000 – General System Characteristics," (CPM report 1999:4, Chalmers University of Technology, Technical Environmental Planning, 2000). See http://msl1.mit.edu/esd123_2001/pdfs/ EPS2000.PDF

www.ingramcontent.com/pod-product-compliance
Lightning Source LLC
Chambersburg PA
CBHW072036190526
45165CB00017B/958